MW01283244

BASIL

Medicinal and Aromatic Plants – Industrial Profiles

Individual volumes in this series provide both industry and academia with in-depth coverage of one major medicinal or aromatic plant of industrial importance.

Edited by Dr Roland Hardman

Volume 1
Valerian
edited by Peter J. Houghton

Volume 2
Perilla
edited by He-Ci Yu, Kenichi Kosuna and Megumi Haga

Volume 3
Poppy
edited by Jenő Bernáth

Volume 4
Cannabis
edited by David T. Brown

Volume 5
Neem
H.S. Puri

Volume 6
Ergot
edited by Vladimír Křen and Ladislav Cvak

Volume 7
Caraway
edited by Éva Németh

Volume 8
Saffron
edited by Moshe Negbi

Volume 9
Tea Tree
edited by Ian Southwell and Robert Lowe

Volume 10
Basil
edited by Raimo Hiltunen and Yvonne Holm

Other volumes in preparation

Allium, edited by K. Chan
Artemisia, edited by C. Wright
Cardamom, edited by P.N. Ravindran and K.J. Madusoodanan
Chamomile, edited by R. Franke and H. Schilcher

Please see the back of this book for other volumes in preparation in Medicinal and Aromatic Plants – Industrial Profiles

BASIL

The Genus *Ocimum*

Edited by

Raimo Hiltunen and Yvonne Holm
Department of Pharmacy, University of Helsinki, Finland

harwood academic publishers
Australia • Canada • China • France • Germany • India • Japan
Luxembourg • Malaysia • The Netherlands • Russia • Singapore
Switzerland

 Printed in Singapore.

Amsteldijk 166
1st Floor
1079 LH Amsterdam
The Netherlands

British Library Cataloguing in Publication Data

A catalogue record for this book is available from the British Library.

ISBN 90-5702-432-2
ISSN 1027-4502

CONTENTS

PREFACE TO THE SERIES

There is increasing interest in industry, academia and the health sciences in medicinal and aromatic plants. In passing from plant production to the eventual product used by the public, many sciences are involved. This series brings together information which is currently scattered through an ever increasing number of journals. Each volume gives an in-depth look at one plant genus, about which an area specialist has assembled information ranging from the production of the plant to market trends and quality control.

Many industries are involved such as forestry, agriculture, chemical, food, flavour, beverage, pharmaceutical, cosmetic and fragrance. The plant raw materials are roots, rhizomes, bulbs, leaves, stems, barks, wood, flowers, fruits and seeds. These yield gums, resins, essential (volatile) oils, fixed oils, waxes, juices, extracts and spices for medicinal and aromatic purposes. All these commodities are traded world-wide. A dealer's market report for an item may say "Drought in the country of origin has forced up prices".

Natural products do not mean safe products and account of this has to be taken by the above industries, which are subject to regulation. For example, a number of plants which are approved for use in medicine must not be used in cosmetic products.

The assessment of safe to use starts with the harvested plant material which has to comply with an official monograph. This may require absence of, or prescribed limits of, radioactive material, heavy metals, aflatoxins, pesticide residue, as well as the required level of active principle. This analytical control is costly and tends to exclude small batches of plant material. Large scale contracted mechanised cultivation with designated seed or plantlets is now preferable.

Today, plant selection is not only for the yield of active principle, but for the plant's ability to overcome disease, climatic stress and the hazards caused by mankind. Such methods as in vitro fertilisation, meristem cultures and somatic embryogenesis are used. The transfer of sections of DNA is giving rise to controversy in the case of some end-uses of the plant material.

Some suppliers of plant raw material are now able to certify that they are supplying organically-farmed medicinal plants, herbs and spices. The Economic Union directive (CVO/EU No 2092/91) details the specifications for the **obligatory** quality controls to be carried out at all stages of production and processing of organic products.

Fascinating plant folklore and ethnopharmacology leads to medicinal potential. Examples are the muscle relaxants based on the arrow poison, curare, from species of *Chondrodendron*, and the antimalarials derived from species of *Cinchona* and *Artemisia*. The methods of detection of pharmacological activity have become increasingly reliable and specific, frequently involving enzymes in bioassays and avoiding the use of laboratory animals. By using bioassay linked fractionation of crude plant juices or extracts, compounds can be specifically targeted which, for example, inhibit blood platelet aggregation, or have antitumour, or antiviral, or any other required activity. With the assistance of robotic devices, all the members of a genus may be readily screened. However, the plant material must be **fully** authenticated by a specialist.

The medicinal traditions of ancient civilisations such as those of China and India have a large armamentarium of plants in their pharmacopoeias which are used throughout South East Asia. A similar situation exists in Africa and South America. Thus, a very high percentage of the world's population relies on medicinal and aromatic plants for their medicine. Western medicine is also responding. Already in Germany all medical practitioners have to pass an examination in phytotherapy before being allowed to practise. It is noticeable that throughout Europe and the USA, medical, pharmacy and health related schools are increasingly offering training in phytotherapy.

Multinational pharmaceutical companies have become less enamoured of the single compound magic bullet cure. The high costs of such ventures and the endless competition from me too compounds from rival companies often discourage the attempt. Independent phytomedicine companies have been very strong in Germany. However, by the end of 1995, eleven (almost all) had been acquired by the multinational pharmaceutical firms, acknowledging the lay public's growing demand for phytomedicines in the Western World.

The business of dietary supplements in the Western World has expanded from the Health Store to the pharmacy. Alternative medicine includes plant based products. Appropriate measures to ensure the quality, safety and efficacy of these either already exist or are being answered by greater legislative control by such bodies as the Food and Drug Administration of the USA and the recently created European Agency for the Evaluation of Medicinal Products, based in London.

In the USA, the Dietary Supplement and Health Education Act of 1994 recognised the class of phytotherapeutic agents derived from medicinal and aromatic plants. Furthermore, under public pressure, the US Congress set up an Office of Alternative Medicine and this office in 1994 assisted the filing of several Investigational New Drug (IND) applications, required for clinical trials of some Chinese herbal preparations. The significance of these applications was that each Chinese preparation involved several plants and yet was handled as a single IND. A demonstration of the contribution to efficacy, of each ingredient of each plant, was not required. This was a major step forward towards more sensible regulations in regard to phytomedicines.

My thanks are due to the staff of Harwood Academic Publishers who have made this series possible and especially to the volume editors and their chapter contributors for the authoritative information.

Roland Hardman

PREFACE

Our interest in the plant named basil began in the early 1980s when the Division of Pharmacognosy at the University of Helsinki was involved in a cultivation project with aromatic plants. Basil is a tender herb, not very suitable for the Finnish climate. It could be successfully cultivated in greenhouse conditions though. Some years later Mr Galambosi brought seeds of different cultivars of *Ocimum basilicum*, cultivated the plants and made us analyse their essential oils. After that we were stuck in the fascinating world of basil oil.

When searching the literature for data on the chemical composition one could not help noticing the confusion in the nomenclature. Thus a need to clarify the taxonomy and nomenclature was born, and through Mr Galambosi we got into contact with Dr Paton, who accepted the task and now provides a delimitation of *Ocimum* from related genera and an infrageneric classification of *Ocimum*.

Chemotaxonomy has been used as a tool to separate the different species. However, it has not been very successful because there are a large number of subspecies, varieties, forms, cultivars and even some hybrids. These are not readily separated on the basis of the essential oil composition. In addition, the essential oil composition is known to vary depending on the cultivation methods, drying and isolation techniques used. The establishment of chemotypes within the species is not easily performed either. There is always one oil which will not fit into an already established group.

Ocimum sanctum or Holy Basil is a sacred medicinal plant in India, where it is used for many different ailments in ethnomedicine. Some activities have been confirmed by pharmacological studies. *O. gratissimum*, *O. viride* and *O. suave* are native plants in different parts of Africa and are used in traditional medicine, mostly as expectorants. The essential oils of these species also exhibit large antimicrobial spectra. An important activity is the insect repelling effect of many *Ocimum* oils, which can be utilized in warm countries. The composition of the fixed oil of *Ocimum* seeds has been studied lately and it was found to have anti-inflammatory activity. In spite of all the pharmacological activities of *Ocimum* essential oils the main use of the plants is as aromatic plants and spices.

This book is intended to cover the present knowledge of all aspects of the cultivation, composition and use of *Ocimum* plants. There certainly are areas which need further exploitation, such as the enantiomeric composition of the essential oils. We hope this book will fulfil your expectations regarding its contents.

Raimo Hiltunen
Yvonne Holm

CONTRIBUTORS

Bertalan Galambosi
Agricultural Research Centre of Finland
Karila Research Station for Ecological Agriculture
Karilantie 2 A
FIN 50600 Mikkeli
Finland

M.M. Harley
Herbarium
Royal Botanic Gardens
Kew
Richmond
Surrey TW9 3AB
UK

R.M. Harley
Herbarium
Royal Botanic Gardens
Kew
Richmond
Surrey TW9 3AB
UK

Raimo Hiltunen
Department of Pharmacy
P.O. Box 56
University of Helsinki
FIN 00014 Helsinki
Finland

Yvonne Holm
Department of Pharmacy
P.O. Box 56
University of Helsinki
FIN 00014 Helsinki
Finland

Seija Marjatta Mäkinen
Department of Applied Chemistry and Microbiology
Division of Nutrition
P.O. Box 27
University of Helsinki
FIN 00014 Helsinki
Finland

Kirsti Kaarina Pääkkönen
Department of Food Technology
P.O. Box 27
University of Helsinki
FIN 00014 Helsinki
Finland

Alan Paton
Herbarium
Royal Botanic Gardens
Kew
Richmond
Surrey TW9 3AB
UK

Eli Putievsky
Agricultural Research Organization
Newe Ya'ar Research Centre
P.O. Box 90000
Haifa 31900
Israel

1. *OCIMUM*: AN OVERVIEW OF CLASSIFICATION AND RELATIONSHIPS

ALAN PATON, M.R. HARLEY and M.M. HARLEY

Herbarium, Royal Botanic Gardens, Kew, Richmond, Surrey TW9 3AB, UK

SUMMARY The taxonomy and nomenclature of *Ocimum* are in a state of confusion. The aim of this paper is to clearly delimit *Ocimum* from related genera, provide an up-to-date infrageneric classification which can be used as a framework for understanding relationships within the genus, provide a list of recognised species with their correct names and common synonyms and a key to their identification. In all 64 species are recognised. A parsimony analysis was carried out which suggests *Ocimum* is a monophyletic group if segregate genera such as *Becium* and *Erythrochlamys* are included within it. An account of the morphological features of *Ocimum* is provided and the economic uses of the genus are briefly discussed in the context of the infrageneric classification.

INTRODUCTION

Ocimum L. is a member of the Labiatae family. The typical characteristics of this family are a square stem, opposite and decussate leaves with many gland dots. The flowers are strongly zygomorphic with two distinct lips. Many of the family, particularly subfamily Nepetoideae, to which *Ocimum* belongs, are strongly aromatic due to essential oils which consist of monoterpenes, sesquiterpenes and phenylpropanoids.

Ocimum, unlike other economically important herbs in the Labiatae such as *Rosmarinus*, *Thymus* and *Salvia*, belongs in tribe *Ocimeae* which has declinate stamens. That is the stamens lie over the lower (anterior) lip of the corolla rather than ascending under the upper (posterior) lip. The *Ocimeae* are essentially a tropical tribe and *Ocimum* occurs naturally in tropical America, Africa and Asia. Unlike several other economic Labiatae, *Ocimum* requires warmth for growth and should be protected from frost.

Ocimum is an important economic and medicinal herb, and yet its taxonomy and nomenclature are in a bit of a muddle. Taxonomy underpins all plant science: we must know the correct name for a plant if we are to communicate information about its uses and relationships. It is important that each species has only one correct name to avoid confusion and aid clarity. The *International Code of Botanical Nomenclature* (Greuter *et al.*, 1994) ensures that names are attached by a standard set of rules. In the literature concerning *Ocimum*, particularly the industrial and economic papers, these rules have frequently not been applied and the same species is often referred to by more than one name.

The circumscription of *Ocimum* itself is also problematic. Estimates of species number vary from 30 (Paton 1992) to 160 (Pushpangadan & Bradu 1995). These differences are due partly to taxonomic reasons, for example, recent revisions such as Paton (1992) have reduced species number by placing some species in synonomy; partly for geographic reasons, in that many species are African and much Indian literature is written without knowledge of the literature pertinent to the African species; and partly because the generic description of *Ocimum* itself has changed. The aim of this paper is to discuss the delimitation of the genus and provide an infrageneric classification of the genus which can be used as a framework for future studies. A list of recognised species is provided which includes information on their distribution, habitats and types in order to help workers use the correct name for a particular species. Some of the more commonly found synonyms are also included under their correct name. A key to the species is provided and references to further information about the species are given.

Ocimum and Related Genera

Briquet (1897) divided Tribe *Ocimeae* into three subtribes. *Ocimum* belongs in Subtribe *Ociminae* characterised by a small, flat lower (anterior) corolla lip with the stamens and style extending over it and up towards the upper (posterior) lip of the corolla. Ryding (1992) divided subtribe *Ociminae* into three informal groups. *Ocimum* was placed in the '*Ocimum*-group' along with *Becium* Lindl., *Erythrochlamys* Gürke, *Hemizygia* (Benth.) Briq., *Syncolostemon* E.Mey. ex Benth. and *Catoferia* (Benth.) Benth. *Orthosiphon* Benth. subgen. *Nautochilus* (Bremek.) Codd must be added to this group as it shares several characters found in *Ocimum* (Paton 1992).

Becium described by Lindley in 1842 contains 35 species, found in Africa, Madagascar and Arabia with one species reaching India. As generic concepts are currently applied, *e.g.* Sebald (1988, 1989), *Becium* can be distinguished from *Ocimum* by having a gland which exudes nectar at the base of the cymes in the inflorescence (Figure 1.1) and by having elongated anthers with parallel thecae rather than orbicular or reniform anthers with diverging thecae. *Becium* usually also has truncate lateral lobes of the calyx, but these are lacking in *B. irvinei* (J.K. Morton) Sebald. When the New World species currently placed in *Ocimum* are examined, this distinction is blurred with several species having a gland at the base of the cymes and divergent anther thecae.

Erythrochlamys was described in 1894 by Gürke and contains two species found in NE Tropical Africa. It has traditionally been separated from *Ocimum* by having an expanded upper lip of the calyx (Baker 1900). However, this character is also seen in *O. circinatum* A.J.Paton (Paton 1992) and *O. transamazonicum* Pereira (Pereira 1972). The character has already been recognized as poor evidence for generic separation with species formerly placed in *Erythrochlamys* by Hedge and Miller (1977) due to the presence of an enlarged upper lip, now being placed in either *Ocimum* (Paton 1992) or *Endostemon* (Paton 1994) on account of similarities in other characters. *E. fruticosus* has appendiculate stamens like many species of *Ocimum*. This again throws doubt on the generic delimitation of *Erythrochlamys*.

Figure 1.1 (a) *Ocimum fimbriatum* showing the bowl-like gland at the base of the cyme; (b) *O. tenuiflorum*; (c) *O. basilicum*; (d) *O. kilimandscharicum*.

Of the other currently recognized genera in the *Ocimum*-group, *Hemizygia* containing 32 species and *Syncolostemon* with 10 species are mainly southern African, but there is one species of *Hemizygia* in Madagascar and one in India. The genera merge into one another and good accounts can be found in Codd (1985). They differ from *Ocimum* by having fused anterior anther filaments. *Orthosiphon* subgenus *Nautochilus* contains 5 species and is found in Southern Africa (Codd 1964). This taxon shares many of the characters of *Ocimum lamiifolium* in particular. *Catoferia* includes 4 species and is found in Tropical America (Ramamoorthy 1986). The genus looks very different from the other species of the *Ocimum* group in having the cyme branches fused to the inflorescence axis, rather than spreading and free, rounded rather than subulate style branches and a bent rather than straight embryo.

Taxonomic History of *Ocimum*

Ocimum was described by Linnaeus in 1753 who listed 5 species. Bentham (1832) recognized just under 40 species and divided *Ocimum* into 3 sections: *Ocimum* [*Ocymodon* Benth.] with appendiculate posterior stamens; *Hierocymum* Benth. with hairs at the base of the posterior stamens and *Gymnocymum* Benth. with glabrous posterior stamens. The latter two sections contained a few species which are now placed in *Endostemon* N.E.Br., a genus Bentham did not recognize, or *Hemizygia.* Bentham (1848) then subdivided section *Ocimum* [*Ocymodon*] into three subsections on the basis of calyx morphology. In subsect. *Ocimum* (*Basilica* sensu Briquet 1897) the throat of the fruiting calyx is open and bearded (Figure 1.2 e–f); in subsect. *Gratissima* the throat is closed by the median lobes of the lower lip being pressed against the under surface of the upper lip (Figure 1.2 c–d, g–h); subsection *Hiantia* Benth., with truncate lateral calyx lobes, only included species which are sometimes placed in *Becium* (Sebald 1988, 1989; Paton 1995; see below). Bentham (1848) also added Sect. *Hemizygia* Benth. which Briquet (1897) considered to be a separate genus on account of the fused anterior stamens. Paton (1992) in his revision of African species of *Ocimum* recognized around 30 species and used Bentham's (1848) infrageneric classification of *Ocimum*, with sect. *Hemizygia* and subsect. *Hiantia* removed, preferring to consider the later as the separate genus *Becium.* This classification is supported by nutlet characters which Bentham did not consider and by analysis of pollen morphology (Harley *et al.*, 1992). However, this classification is not entirely without problems, as pointed out (Paton 1992) *Ocimum circinatum* A.J. Paton does not fit neatly into the existing categories. *Ocimum lamiifolium* is also anomalous, appearing to have a close relationship to *Orthosiphon* subgenus *Nautochilus* (Bremek.) Codd.

Pushpangadan (1974; Pushpangadan & Bradu 1995; Sobti & Pushpangadan 1979) formulated a different infrageneric classification. The '*Basilicum*' group contains herbaceous annuals or sometimes perennials with black, ellipsoid, strongly mucilaginous seeds and with a basic chromosome number of n = 12, whereas the '*Sanctum*' group are perennial shrubs with brown globose non-mucilaginous or weakly mucilaginous seeds and a basic chromosome number of n = 8. The *Basilicum* group contains only section *Ocimum* subsection *Ocimum.* The remainder of the genus must be placed in the *Sanctum* group. This classification is commonly used in the economic and industrial literature, *e.g.* Darrah (1980), Pushpangadan and Bradu (1995) whereas Bentham's system is the basis for that used in taxonomic literature.

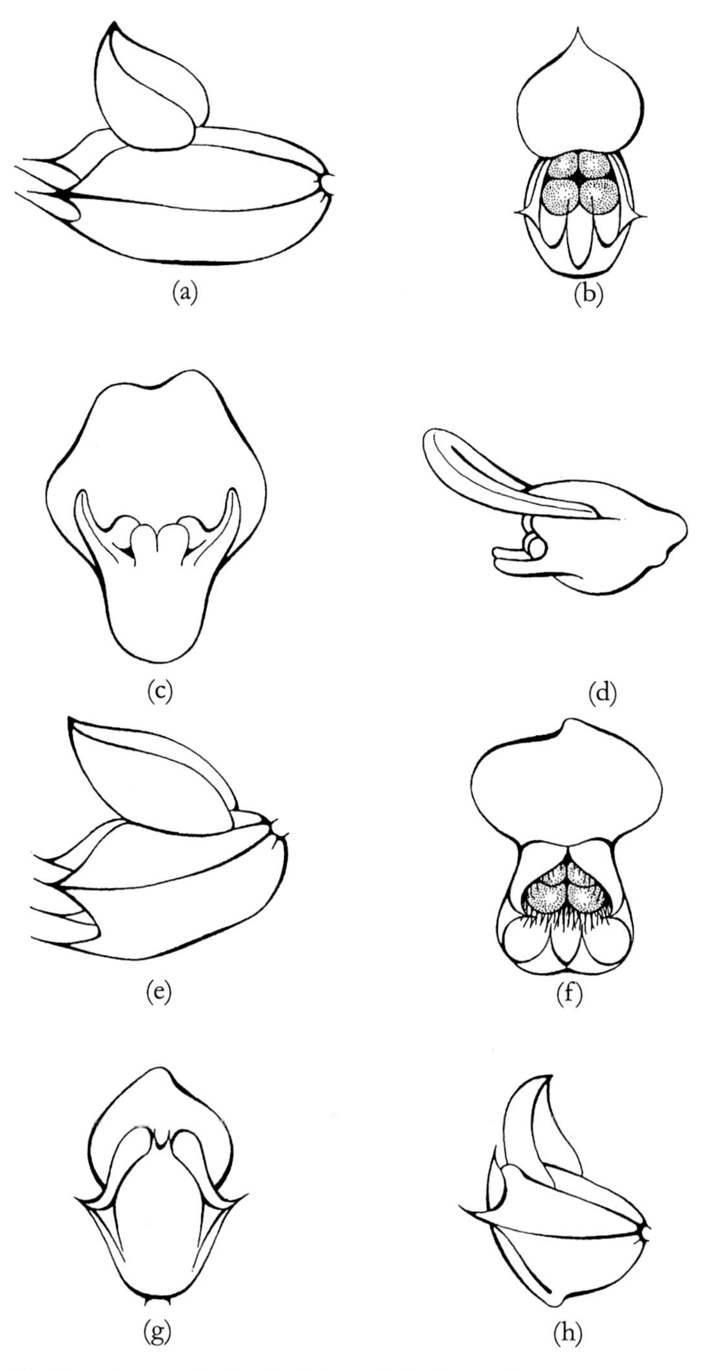

Figure 1.2 Fruiting calyx morphology in *Ocimum*. (a–b) *Ocimum lamiifolium*. (a) side view; (b) front view. (c–d). *O. cufodontii*. (c) view from beneath; (d) side view. (e–f) *O. basilicum* (e) side view; (f) front view. (g–h) *O. gratissimum*. (g) view from beneath; (h) side view. Drawn by Emmanuel Papadopoulos. Reproduced from Paton (1992) by permission of RBG Kew.

There are several problems with Pushpangadan's infrageneric classification. It does not adequately convey the variation within the genus. New World and Old World species which differ considerably in morphology are placed together within the *Sanctum* group. However, within the *Sanctum* group there are several distinct groups of species with similar attributes and this information is lost. Pushpangadan's system does not comply with the *International Code of Botanical Nomenclature* and thus should not be used as a standard.

The most recent study of New World species of *Ocimum* is that of Epling (1936) who recognizes 12 species 4 of which are pantropical, widely cultivated species. Epling does not refer to the infrageneric classification of the genus.

GENERIC AND INFRAGENERIC DELIMITATION OF *OCIMUM*

Currently there is no up-to-date infrageneric classification which adequately communicates the variation of *Ocimum* throughout its geographical range, nor clearly indicates how the genus should be delimited from its closest allies. A parsimony analysis of the genus and close relatives was carried out to try and determine the generic delimitation and infrageneric relationships and to produce a classification which could serve as a standard foundation for the further scientific and economic investigation of *Ocimum*. The parsimony analysis of the *Ocimum*-group of genera was carried out using PAUP version 3.1.1 (Swofford 1993) run on a Power MacIntosh 8100/80.

The choice of sample used for the parsimony analysis was based on a herbarium study of the whole of *Ocimum*, *Becium*, *Erythrochlamys* and *Orthosiphon* subgenus *Nautochilus*. Although this group contains some 64 species many of the species were similar when only conservative characters were examined. For example within subsect. *Ocimum*, Paton (1992) recognized 6 species. However, they differ only in habit and leaf shape, characters which are very variable phylogenetically. A similar situation applies in *Becium*. This may indicate that speciation within the currently recognized *Ocimum* and *Becium* in Africa has been fairly recent. In all, a representative sample of 20 taxa were chosen to represent the variation in *Ocimum*, *Becium*, *Erythrochlamys*, and *Orthosiphon* subgenus *Nautochilus*. All Bentham's (1848) sections and subsections of *Ocimum* were represented as follows:

1. Sect *Ocimum* subsect. *Ocimum*. This section contains 7 species and is represented in the analysis by *O. americanum*. The other species in the group only differ in leaf morphology and habit.
2. Sect. *Ocimum* subsect. *Gratissima*. This section contains 6 species and is represented in the analysis by *O. gratissimum*, *O. jamesii*, *O. cufodontii*.
3. Sect. *Ocimum* subsect. *Hiantia* (sometimes recognized as *Becium*). This group contains 35 species and is represented by *B. fimbriatum*, *B. grandiflorum*, *B. dhofarense* and *B. irvinei*. This represents one species from each of the infrageneric groups of *Becium* recognized by Sebald (1988, 1989). There is little morphological variation within Sebald's groups.
4. Sect. *Hierocymum*. Bentham recognised 11 species in this group, but some have been moved to *Endostemon* (Paton 1994) or *Platostoma* (Paton 1997). The group is

represented in this study by *O. lamiifolium* Benth., *O. selloi* Benth., *O. nudicaule* Benth. and *O. tenuiflorum* L.

5. Sect. *Gymnocimum*. Bentham recognized 8 species in this group. Most have been moved to other genera such as *Fuerstia* (Paton 1995b), *Endostemon* (Paton 1994) or *Hemizygia* (Codd 1985). The group is represented in the analysis by *O. campechianum* Mill. and *O. ovatum* Benth.
6. *O. circinatum* is also included in the sample.

Erythrochlamys is represented by both its species *E. fruticosus* and *E. spectabilis*. *Orthosiphon* subgen. *Nautochilus* includes 5 species (Codd 1985) and is represented by *O. labiatus* and *O. tubiformis*. *Syncolostemon*, *Hemizygia* and *Catoferia* were also included in order to investigate the generic delimitation of *Ocimum*. *Orthosiphon* subgenus *Orthosiphon* from the *Orthosiphon* group (Ryding 1994) was chosen as the outgroup and the oligomorphic genera *Syncolostemon*, *Hemizygia* and *Catoferia* were scored as single taxa. *Orthosiphon aristatus* (Blume) Miq. was added because of its resemblance to *Catoferia* (Hedge 1992).

Morphology

Morphological variation within *Ocimum* and allies is summarized below. Characters used in the cladistic analysis can be found under the appropriate section. The number in parenthesis preceding the character refers to that character's position in the data matrix. The data matrix used for the analysis is provided in Appendix 1.

1. Habit

Members of the genus can be annual herbs which lack a rhizome, possessing only a tap root (*e.g.*, *O. basilicum* L.) or they can be suffrutices with a large or small perennating rhizome which can produce annual or short-lived stems (*e.g.*, *O. obovatum* E.Mey. ex Benth.). Sometimes they are shrubs with quite woody stems (*e.g.*, *O. forskolei* Benth.) or they can be much softer subshrubs (*e.g.*, *O. tenuiflorum*). The habit is very variable with closely related species having very different habits. For example, on evidence from other characters, the species of *Ocimum* sect. *Ocimum*, subsect. *Ocimum* (as defined by Paton 1992) are closely related, but annuals, shrubs and suffrutices occur within this group. The whole plants are often strongly aromatic, but some species have a very weak scent, *e.g.*, *O. filamentosum* Forssk. Habit characters were not used in cladistic analysis.

2. Leaf scars

The leaves can be petiolate or sessile, often toothed at the margin. Some species from arid regions, *e.g.*, *O. cufodontii* (Lanza) Paton, *Erythrochlamys spectabilis* Gürke have deciduous leaves which leave behind conspicuous leaf scars.

(3) **Leaf scars**: absent (0); present (1).

3. Inflorescence

The typical inflorescence in *Ocimum* is a thyrse composed of opposite 1–3-flowered cymes. Two opposing cymes are sometimes called a verticil and several verticils are usually borne on the inflorescence axis. Cymes are usually 3-flowered but rarely, *e.g.*, in *O. irvinei* J.K. Morton, they are reduced to a single flower. The cymes are subtended by bracts which sometimes abscise early, *e.g.*, sect. *Hiantia*. In these cases the bract scar often develops into a bowl-shaped, gland-like structure which serves as an auxiliary nectary and is frequently visited by ants (Figure 1.1a) (Vogel 1998). In *Catoferia* the cyme axis is fused to the inflorescence axis giving the appearance that the flowers arise several millimetres above the bract. In all other species studied the cyme was sessile and the flower pedicel arose from immediately above the bract.

(1) **Bract persistence:** persistent (0); caducous (1).
(2) **Bract abscission gland**: absent (0); present (1).
(4) **Cyme flower number:** 3-flowered (0); 1-flowered (1); variable 1–3-flowered (2).
(5) **Cyme branches:** free, not fused to stem (0); fused to stem (1).

4. Calyx

The calyx is generally shortly tubular or funnel-shaped; it is straight or slightly curved, but in *O. transamazonicum* it is bent down forming a right angle at the throat. The calyx is 5-lobed with an upper (posterior) single-lobed lip and a 4-lobed lower (anterior) lip. The anterior lip consists of 2 lateral and 2 median lobes. The calyx usually enlarges slightly in fruit. The posterior lip in *Ocimum* is decurrent on the tube forming inconspicuous or prominent wings. In some species, *e.g. O. cufodontii* (Lanza) A.J.Paton, *O. transamazonicum* and *E. spectabilis* the posterior lip becomes membranous and expanded, forming a sail-like wing. In these species the whole calyx containing the ripe nutlets is often dispersed, rather than the nutlets themselves.

The throat of the calyx can be open and glabrous, e.g., *O. lamiifolium* (Figure 1.2a–b), or open and bearded with a ring of hairs just below the mouth which is most noticeable at the fruiting stage, *e.g.*, *O. basilicum* (Figure 1.2e–f). In *E. spectabilis* for example, and in subsect. *Gratissima*, the median lobes of the anterior lip close the calyx throat by pressing against the posterior lip (Figure 1.2c–d, g–h). In some species *e.g.*, subsect. *Hiantia* the tube is slightly constricted above the nutlets in fruit making the mouth laterally compressed.

The shape and position of the lobes of the anterior lip also vary. The lateral lobes can be lanceolate and symmetrical as in *O. basilicum.* They can be asymmetric but still with a lanceolate tooth, *e.g.*, *O. gratissimum* and *O. lamiifolium.* In these cases the posterior margin of the lateral lobe is extended towards the posterior lip forming a distinct shoulder (Figure 1.2). In *Catoferia chiapensis* the lateral lobes are fused to the median lobes, the division becoming indistinct. In some species of subsect. *Hiantia* the lateral lobe is truncate and not separated from the median lobes by a distinct sinus. In this case the median lobes are small and subulate, in most other species they are lanceolate, but in *e.g.*, subsect. *Gratissima,* the median lobes are fused for some of their length giving an emarginate appearance (Figure 1.2). The lateral lobes are usually situated midway between the median lobes of the anterior lip and the posterior lip, but in *Orthosiphon aristatus* and *Catoferia* the lateral lobes are much closer to the median lobes than the posterior lip.

(6) **Hairy annulus in calyx throat:** absent (0); present (1).
(7) **Posterior lobe:** decurrent (0); not decurrent (1).
(8) **Posterior lobe:** not expanded into sail (0) expanded (1).
(9) **Posterior lobe:** not membranous (0); membranous (1).
(10) **Lateral lobe position:** midway (0); nearer median lobes (1).
(11) **Lateral lobe:** lanceolate symmetric (0); truncate (1); lanceolate asymmetric (2); indistinct (3).
(12) **Median lobes:** lanceolate (0); emarginate (1); subulate (2).
(13) **Tube throat**: median lobes of anterior lip not closing tube (0); closed by median lobes of anterior lip (1).
(14) **Tube (lateral constriction):** open (0); constricted by lateral lobes (1).
(15) **Calyx curve:** straight (0); bent down at mouth (1).

5. Corolla

The corolla is usually straight, but can be curved downwards slightly. The tube is often dorsally gibbous at the mid point, at a point opposite the appendage of the posterior stamen in species which display this, *e.g.*, *O. basilicum*. Some species which lack an appendage still have a gibbous corolla tube, *e.g.*, *O. lamiifolium*. The tube usually dilates towards the mouth, but in *O. lamiifolium* and allies the tube is parallel-sided towards the mouth. The posterior lip of the corolla is always 4-lobed, but the lobes can be equal as in *e.g.*, *O. basilicum* or the 2 median lobes can exceed the lateral lobes as in *e.g.*, *O. obovatum*. The anterior lip is entire and usually horizontal, but sometimes becomes deflexed.

(16) **Corolla tube base**: not gibbous (0); gibbous (1).
(17) **Corolla tube**: dilating towards mouth (0); not dilating towards mouth (1).
(18) **Posterior lip**: median lobes exceeding lateral (0); equally 4-lobed (1).

6. Androecium

There are always 4 stamens, an anterior pair which attach near the corolla mouth and a posterior pair which attach near the corolla base. In some related genera, *e.g.*, *Orthosiphon* subgenus *Orthosiphon*, the posterior pair attach to the corolla at the mid-point of the tube. In *O. campechianum* and *O. tenuiflorum* the corolla is very small and the position of attachment is difficult to compare to that of other species. Although in absolute terms the stamens attach near the base of the tube, they are situated at the mid-point of the tube as they are more or less equidistant between the base and mouth. The posterior stamen can be straight and lacking an appendage, *e.g.*, *O. ovatum* and *O. tenuiflorum*; basally bent with no unilateral swelling, *e.g.*, *O. lamiifolium*; basally bent with a much reduced appendage, *e.g.*, *O. selloi*; basally bent with a conspicuous appendage, *e.g.*, *O. basilicum*. The bases of the posterior stamens are often pubescent, but sometimes glabrous as in *O. campechianum* and *O. ovatum*. All stamens are fertile with the exception of *O. circinatum* which has infertile posterior stamens.

The anterior stamens are usually free, but in *Hemizygia* and *Syncolostemon*, the anterior pair are fused together.

The anther is two locular, synthecous and dorsifixed. The locules can be parallel, *e.g.*, *O. obovatum* or divergent, *e.g.*, *O. basilicum*. In *O. ovatum* and *O. campechianum* the anther locules are unequal in length, whereas they are equal in all other species examined.

(19) **Posterior stamen appendage:** absent (0); with an inconspicuous appendage (1); with a conspicuous appendage (2).
(20) **Posterior stamen form:** straight (0); bent (1).
(21) **Posterior stamen indumentum:** glabrous (0); basally pubescent (1).
(22) **Posterior stamen attachment:** midpoint of tube (0); base of tube (1); uncertain (?).
(23) **Anther:** locules divergent (0); locules parallel (1).
(24) **Locules:** equal (0); unequal (1).
(27) **Anterior stamens:** free (0); fused together (1).

7. Gynoecium and Disk

The ovary in all species is divided into 4. These parts develop into single seeded nutlets or mericarps. The nutlets can be spherical as in, *e.g.*, *O. gratissimum* or elliptic as in, *e.g.*, *O. basilicum*. They are glabrous, but can be rarely pubescent, *e.g.*, *O. cufodontii.* The nutlets can produce copious mucilage when wet, *e.g.*, *O. basilicum*, a small amount of mucilage, *e.g.*, *O. gratissimum*, or no mucilage at all, *e.g.*, *O. lamiifolium.* The nutlets of *Hemizygia* and *Syncolostemon* have a conspicuous vein on the side closest to the calyx. Ryding (1992) provides a good account of the details of nutlet microstructure.

The style is gynobasic. The apex is bifid with subulate lobes in *Ocimum*, *Erythrochlamys*, *Orthosiphon* subgenus *Nautochilus, Synclostemon* and *Hemizygia*, but bifid with rounded lobes in *Orthosiphon* subgenus *Orthosiphon* and *Catoferia.*

The disk is either equally 4-lobed as in most species of *Ocimum* or the anterior lobe can be larger as in *Orthosiphon* subgenus *Orthosiphon.*

(25) **Style:** lobes rounded (0); lobes bifid (1).
(26) **Disk:** anterior lobe larger (0); equal lobed (1).
(28) **Nutlet shape:** ellipsoid (0); spherical (1).
(29) **Nutlet indumentum:** glabrous (0); pubescent (1).
(30) **Nutlet veins:** veins absent or very faint (0); conspicuous vein on the side closest to the calyx (1)

8. Pollen

The pollen of *Ocimum* and relations is described in detail in Harley *et al.* (1992); Paton (1994).

(31) **Polar outline:** circular (0); ellipsoid (1).
(32) **Intersections of muri of primary reticulum:** rounded (0); angled (1).
(33) **Lacunae formed from primary reticulum muri:** ± isodiametric in mesocolpia (0); elongated in mesocolpia (1).
(34) **Secondary reticulum:** not suspended, supported by columellae (0); suspended, lace-like, not supported by columellae (1).
(35) **Colpus border:** absent (0); present (1).
(36) **Zone between colpi:** equal (0); alternately wide and narrow (1).

Results of Parsimony Analysis

The analysis was carried out using the branch and bound option provided by PAUP which identifies all the most parsimonious trees. The options MULPARS on and addition sequence simple were employed (Swofford 1993). All characters were treated as unordered. 8 trees were found of length = 84, CI = 0. 631, RI = 0. 785. The strict consensus tree is shown in Figure 1.3. Figure 1.4 shows one of the most resolved trees with characters optimized using the "acctran" option of PAUP (Swofford 1993).

It is important that genera can be described unambiguously. Genera should also reflect phylogeny. Monophyletic groups (i.e. ones which contain all the descendants of an ancestor) are desirable as the resultant classification will accurately reflect the hierarchy which is the natural outcome of the evolutionary process (Donoghue & Cantino 1988). From Figure 1.3 it can be seen that *Ocimum* is only monophyletic if *Erythrochlamys*, *Becium* and *Orthosiphon* subgen. *Nautochilus* are incorporated into it. Recognition of any of these taxa at generic rank would render *Ocimum* a paraphyletic group, that is one which only contained some of the descendants of an ancestor. Applied biologists always criticise taxonomists for changing the currently accepted names of plants. However, maintaining the *status quo* is not an option in this case as a lot of information on relationships would be lost.

Clade N (E African *Becium*) (Figure 1.3) is the most robust clade, being present on all the cladograms up to three steps longer than the shortest. All other clades were not indicated on the strict consensus tree of all trees one step longer than the most parsimonious tree. However, it is difficult to clearly delimitate clade N from the rest of *Ocimum*. There is a continuum of variation from New World species (clade K) such as *O. selloi* which have glands at the base of their cymes and divergent anther thecae; through *B. irvinei* which has glands and parallel anther thecae, but still has a calyx similar to *O. selloi*; to E African species of *Becium* (clade N) which have glands and parallel thecae but usually have a truncate lateral lobe to the calyx. Arbitrarily dividing this continuum would not allow the classification to reflect this path of evolution and similarity. There may be similarities in other characters such as chemistry which could be useful to the applied biologist. Therefore *Becium* is not recognized as a separate genus here.

Recognition of *Erythrochlamys* as a separate genus creates other problems. Figure 1.4 demonstrates that this genus could only be diagnosed by the calyx upper lip being expanded and membranous and in having elongated primary muri of the pollen. The pollen character is variable within *E. fruticosus* and occurs in other species and the calyx characters are also found elsewhere. This makes it very difficult to unambiguously describe *Erythrochlamys* and generic recognition would obscure the similarity between these species and clade J (Figure 1.3). Similar problems would occur if *Orthosiphon* subgenus *Nautochilus* were recognised at generic rank as *Ocimum lamiifolium* shares some but not all of the characters of that group.

Bentham (1848) placed *Hemizygia* within *Ocimum*, but recognized *Syncolostemon* as a separate genus. The analysis clearly supports a close relationship between *Syncolostemon* and *Hemizygia* so the question is whether both genera be included within *Ocimum* or not. The *Syncolostemon*/*Hemizygia* complex is monophyletic and the sister group to an enlarged monophyletic *Ocimum*. The *Syncolostemon*/*Hemizygia* complex and *Ocimum*

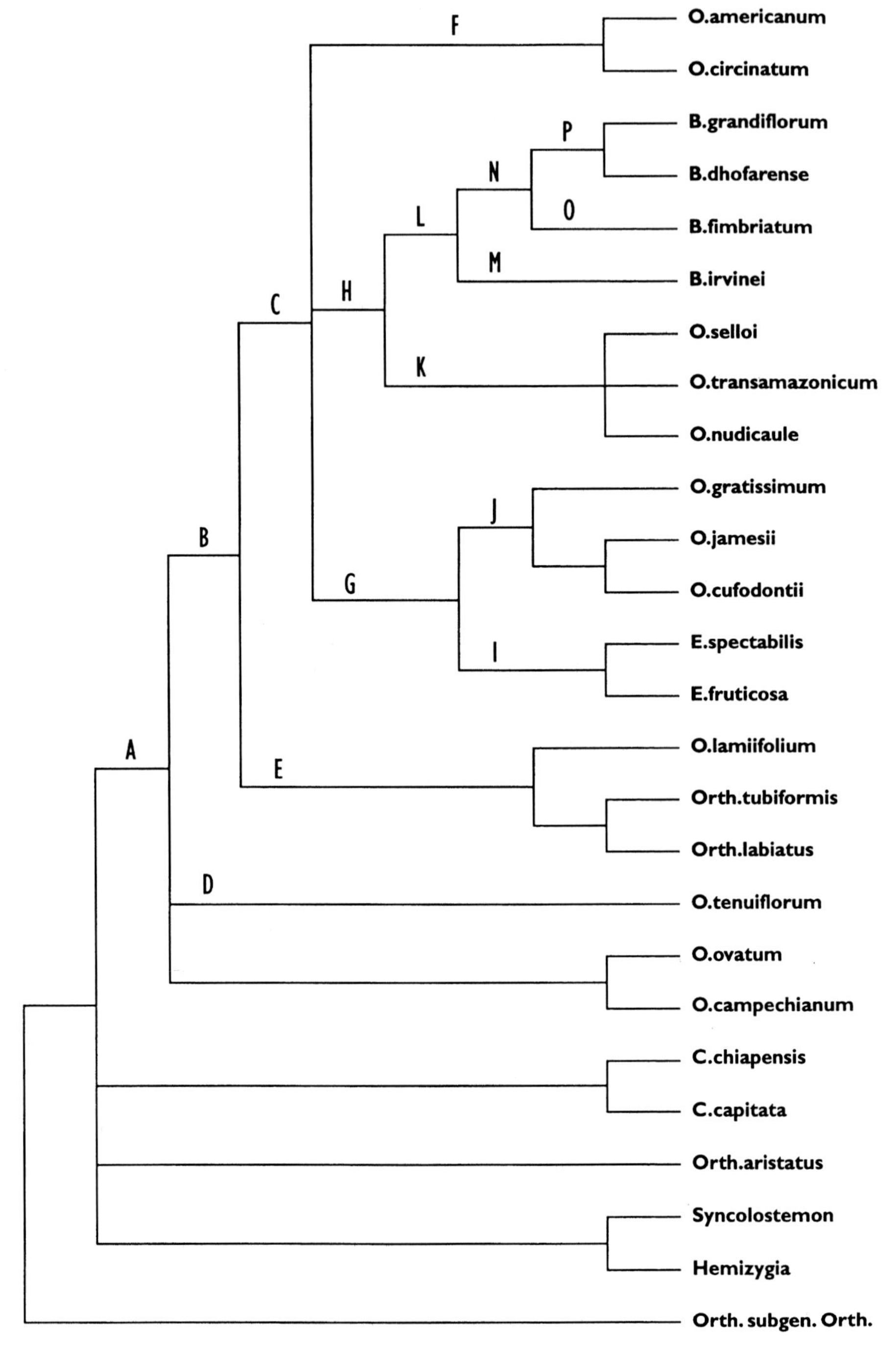

Figure 1.3 The strict consensus tree of 8 trees found in analysis of length = 84, CI = 0. 631, RI = 0. 785. Numbers refer to the parsimony decay values, letters correspond to clades refered to in the text.

differ from *Catoferia* and the remainder of *Orthosiphon* by having basally attached posterior staminal filaments rather than filaments which attach to the midpoint of the corolla tube, and by having subulate style branches rather than rounded style branches. *Ocimum* differs from the *Syncolostemon/Hemizygia* complex by having an equally lobed rather than unequally lobed disk and in having free rather than fused anterior filaments. The nutlets of *Syncolostemon* and *Hemizygia* have a prominent dorsal vein which is lacking in *Ocimum*. There are also several characters of pollen morphology which support this delimitation. As there is no problem in unambiguously describing *Ocimum* and the *Syncolostemon/Hemizygia* complex and as both are monophyletic there is no reason to upset the currently recognized circumscription any further. The question whether *Syncolostemon* and *Hemizygia* should be recognized as separate genera is beyond the scope of this paper, but serious consideration should be given to this problem, basing a study on the whole range of the complex, including the Madagascan and Indian relatives, rather than just the Southern African material considered by Codd (1985) in his treatment of the complex.

In conclusion, the genus *Ocimum* incorporating the previously recognized taxa *Becium*, *Erythrochlamys* and *Orthosiphon* subgenus *Nautochilus*, is easy to describe and communicate. *Ocimum* defined as above is also a monophyletic group (Clade A Figure 1.3) which is useful if the occurrence of attributes is to be predicted.

Infrageneric Classification of *Ocimum*

The strict consensus tree (Figure 1.3) broadly supports the major groupings recognized by Bentham (1848) as follows: sect. *Gymnocimum* (clade D); sect *Ocimum* (clade C); subsect. *Ocimum* (clade F); subsect *Gratissima* (clade J); subsect *Hiantia* (clade M). However, sect. *Hierocymum* is not supported and is polyphyletic on the basis of this analysis, its members being scattered throughout several clades (*O. lamiifolium*, *O. selloi*, *O. nudicaule* Benth. and *O. tenuiflorum* L.).

Bentham's classification does not demonstrate the relationships between *Erythrochlamys* (clade I) and subsect. *Gratissima* (clade J); nor between clade K and clade L nor between *Orthosiphon* subgenus *Nautochilus* and *O. lamiifolium*. This would be possible by formally recognizing clade H, clade G and clade E (Figure 1.3). Bentham's classification is not complex enough to communicate the relationships within *Ocimum*. The following new classification, based on monophyletic groups is an attempt to provide a classification which illustrates relationships and provides readily communicable groups. The infrageneric taxa are listed below with their types, basionyms, and diagnosing features. The full distribution of character states is illustrated in Figure 1.4.

Ocimum L., Sp. Pl.: 597 (1753); Gen. Pl. ed. 5: 259 (1754). Lectotype species: *O. basilicum.*

Becium Lindl. in Edwards's Bot. Reg. 28, Misc.: 42 (1842). Type species: *B. grandiflorum* Lam.
Erythrochlamys Gürke in Bot. Jahrb. Syst. 19: 222 (1894). Type species *E. spectabilis Gürke*
Nautochilus Brem. in Ann. Transv. Mus. 15: 253 (1933).

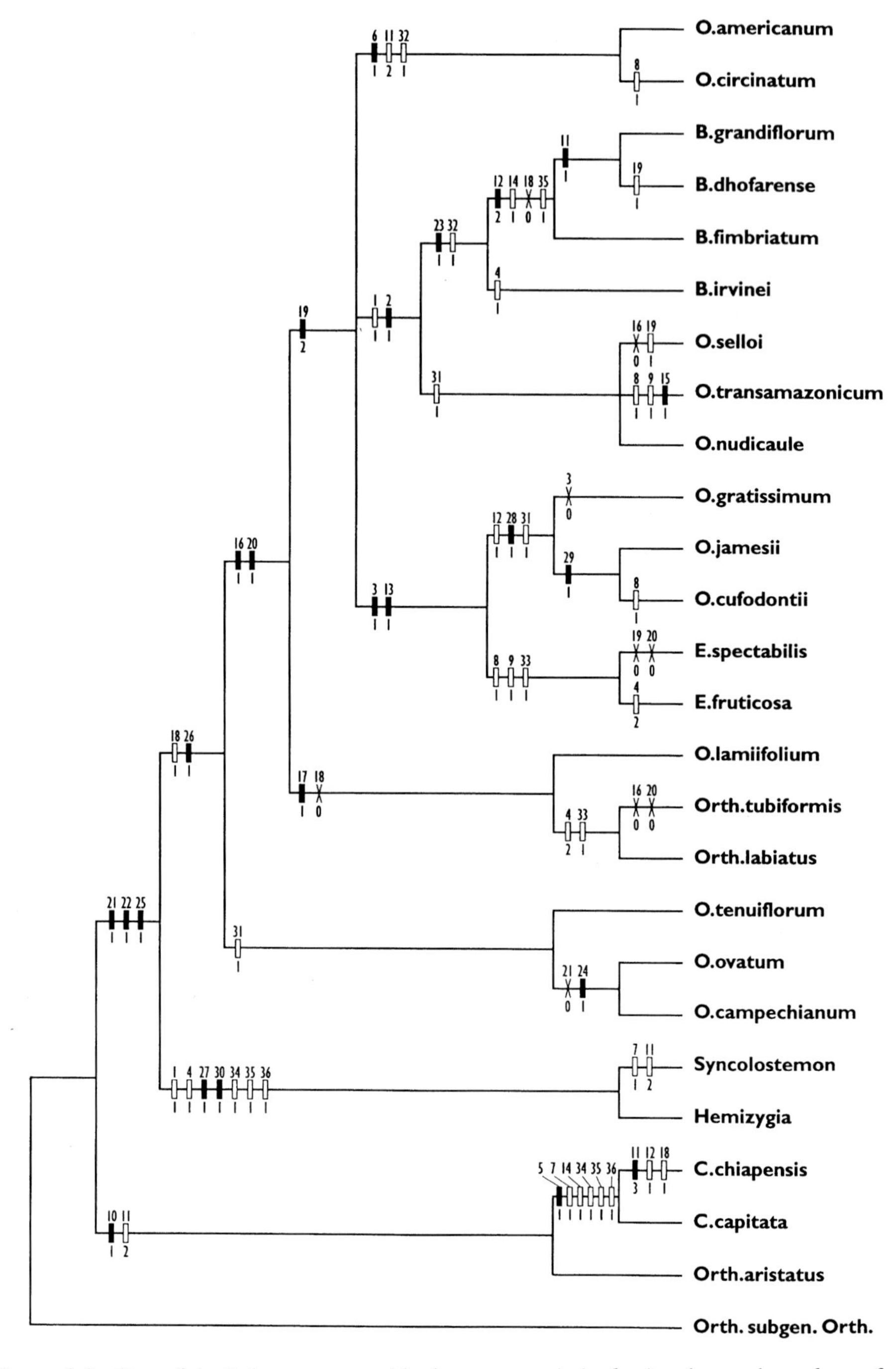

Figure 1.4 One of the 8 shortest trees with characters optimized using the accelerated transformation option of PAUP (Swofford 1993). Numbers above the symbols refer to the character number, numbers below refer to the character state evolved. Solid bars represent unique character changes, outlines bars represent parallel changes, crosses represent character reversals.

Orthosiphon sect. *Serrati* Ashby in Journ. Bot. 76: 1 (1938).
Orthosiphon subgenus *Nautochilus* (Brem.) L.E.Codd in Bothalia 8: 150 (1964). Type species *O. labiatus* (N.E.Br.) Brem.

1. subgenus **Ocimum** (Figure 1.3, Clade C, species 1–54 in enumeration). Type species: *O. basilicum* L.
sect. *Ocymodon* Benth., Labiat. Gen. Sp. 1: 2 (1832)
Corolla tube usually gibbous at midpoint, dilating towards mouth; posterior lip regularly 4-lobed or with median lobes exceeding lateral lobes. Posterior stamens usually bent and appendiculate, very rarely straight and inappendiculate, basally pubescent.

1a. section **Ocimum** (Figure 1.3, Clade F, species 1–9 in enumeration).Type species: *O. basilicum L.*
sect. *Ocimum* subsect. *Ocimum* in Paton, Kew Bull. 47: 421 (1992).
subsect. *Basilica* Briq. in Engl. & Prantl, Nat. Pflanzenfam. 4,3a: 369 (1897).
Bracts persistent. Calyx throat open in fruit; throat with a hairy annulus; lateral lobes lanceolate, symmetrical. Pollen with angled muri of primary reticulum.

1b. section **Gratissima** *(Benth.) A.J. Paton* **stat. nov**. (Figure 1.3, Clade G, species 10–17 in enumeration). Type species: *O. gratissimum* L.
Basionym: subsect. *Gratissima Benth.* in DC. Prod. 12: 34 (1848).
Bracts persistent. Calyx throat closed in fruit by median lobes of lower lip pressing towards upper lip; throat glabrous; lateral lobes asymmetric. Pollen with rounded muri of primary reticulum.

1b. I. subsection **Gratissima** *Benth.* in DC. Prod. 12: 34 (1848). (Figure 1.3, Clade J, species 10–15 in enumeration).
Posterior lobe of calyx not membranous, expanded or not. Median lobes of anterior lip of calyx denticulate. Nutlets spherical. Pollen with an ellipsoid polar outline with isodiametric lacunae formed by primary muri.

1b. II. subsection **Erythrochlamys** *(Gürke) A.J. Paton* **comb. nov.** Type species: *O. spectabilis* (Gürke) A.J.Paton. (Figure 1.3, Clade I, species 16–17 in enumeration).
Erythrochlamys Gürke in Bot. Jahrb. Syst. 19: 222 (1894).
Posterior lobe of calyx membranous, expanded. Median lobes of anterior lip of calyx lanceolate. Nutlets obovate. Pollen with a circular polar outline with elongate lacunae formed by primary muri.

1c. section **Hiantia** *(Benth.) A.J. Paton* **stat. nov**. (Figure 1.3, Clade H, species 18–54 in enumeration). Type species: *O. filamentosum* Forssk.
Basionym: *Ocimum* subsect. *Hiantia* Benth. in DC. Prodr. 12: 35 (1848).
Bracts caducous, scar developing into an auxiliary nectary. Calyx throat open or laterally compressed in fruit; throat glabrous; lateral lobes asymmetrically lanceolate or truncate. Pollen with rounded or angled muri of primary reticulum.

1c. I. subsection **Hiantia** *Benth* in DC. Prodr. 12: 35 (1848). (Figure 1.3, Clade L, species 21–54 in enumeration) Lectotype species chosen here: *O. filamentosum* Forssk.

Becium Lindl. in Edwards's Bot. Reg. 28, Misc.: 42 (1842). Type species: *B. grandiflorum* Lam.

Anthers elongate with parallel thecae. Pollen with circular polar outline and with angled muri of primary reticulum.

1c. I. a. series **Monobecium** (*Sebald*) *A.J. Paton* **comb. & stat. nov**. (Figure 1.3, Clade M, species 21 in enumeration). Type species: *Ocimum irvinei* J.K. Morton.

Basionym: *Becium* subgenus *Monobecium* Sebald in Stuttgarter Beitr. Naturk. A, 419: 34 (1988).

Cymes 1-flowered. Lateral lobes of calyx asymmetrically lanceolate with a sinus between lateral and median lobes; median lobes of anterior lip lanceolate. Posterior lip of corolla regularly 4-lobed. Stamens appendiculate. Pollen lacking a colpal border of primary muri.

1c. I. b. series Serpyllifolium (*Sebald*) *A.J. Paton* **comb. & stat. nov**. (Figure 1.3, Clade O, species 22–26 in enumeration).Type species: *O. serpyllifolium* Forssk.

Basionym: *Becium* subgen. *Becium* sect. *Serpyllifolium* Sebald in Stuttgarter Beitr. Naturk. A, 405: 10 (1987).

Cymes 3-flowered. Lateral lobes of calyx asymmetrically lanceolate with sinus between lateral and median lobes; median lobes of anterior lip subulate. Posterior lip of the corolla with median lobes exceeding lateral lobes. Stamens appendiculate or rarely inappendiculate. Pollen with a colpal border of primary muri.

1c. I. c. series **Hiantia** (*Benth.*) *A.J. Paton* (Figure 1.3, Clade P, species 27–54 in enumeration). Lectotype species chosen here: *O. filamentosum* Forssk.

Basionym: subsect. *Hiantia* Benth in DC. Prodr. 12: 35 (1848).

Becium subgen. *Becium* sect. *Becium* in Sebald, Stuttgarter Beitr. Naturk. A, 419: 51 (1988).

Becium subgenus *Orthobecium* Sebald in Stuttgarter Beitr. Naturk. A, 405: 10 (1987).

Cymes 3-flowered. Lateral lobes of calyx truncate with no sinus between lateral and median lobes; median lobes of anterior lip subulate. Posterior lip of the corolla with median lobes exceeding lateral lobes. Stamens appendiculate or rarely inappendiculate. Pollen with a colpal border of primary muri.

1c. II. subsection **Nudicaulia** (**Briq**.) ***A.J.Paton*** **comb. nov.** (Figure 1.3, Clade K, species 18–20 in enumeration). Type species: *O. nudicaule.*

Basionym: Sect. *Hierocimum* subsect. *Nudicaulia* Briq. in Engl. & Prantl, Nat. Pflanzenfam. 4,3a: 372 (1897).

Anthers rotund to reniform with divergent thecae. Pollen with ellipsoid polar outline and with rounded muri of primary reticulum.

2. subgenus **Nautochilus** (*Bremek.*) *A.J.Paton* **comb. nov.** Type species: *Ocimum labiatus* (N.E.Br.) A.J. Paton. (Figure 1.3, Clade E, species 55–60 in enumeration).

Basionym: *Nautochilus* Bremek. in Ann. Transv. Mus. 15: 253 (1933).
Orthosiphon sect. *Serrati* Ashby in J. Bot., London 76: 1 (1936)
Orthosiphon subgenus *Nautochilus* (Bremek.) Codd in Bothalia 8: 150 (1964).
Corolla tube gibbous or not at midpoint, parallel-sided distally, not dilating towards mouth; posterior lip with median lobes exceeding lateral lobes. Posterior stamen straight or basally bent, inappendiculate, basally pubescent.

3. subgenus **Gymnocimum** (*Benth.*) *A.J. Paton* **stat. nov.** (Figure 1.3, Clade D plus *O. tenuiflorum*, species 61–64 in enumeration). Type species: *O. campechianum* Miller.
Basionym: sect. *Gymnocimum* Benth., Labiat. Gen. Sp.: 12 (1832).
Corolla tube not gibbous at base, dilating towards mouth; posterior lip regularly 4-lobed. Posterior stamens straight, inappendiculate, glabrous or basally pubescent.

3. I. sect. **Gymnocimum** *Benth.* (Figure 1.3, Clade D, species 61–63 in enumeration). Lectotype species chosen here: *O. campechianum* Miller.
Sect. *Gymnocimum* subsect. *Brevituba* Briq. in Engl. & Prantl, Nat. Pflanzenfam. 4,3a: 371 (1897).
Posterior stamens glabrous. Anther thecae unequal.

3. II. sect. **Hierocymum** *Benth.* (species 64 in enumeration). Lectotype species chosen here: *O. tenuiflorum L.*
Sect. *Hierocimum* subsect. *Foliosa* Briq. in Engl. & Prantl, Nat. Pflanzenfam. 4,3a: 371 (1897). Lectotype species chosen here: *O. tenuiflorum.*
Posterior stamens basally pubescent. Anther thecae equal.
The relationships within subgenus *Gymnocimum* are still obscure, the position of *O. tenuiflorum* being uncertain with some of the 8 trees placing it at the base of clade D and others in an unresolved position.

Key to infrageneric taxa within *Ocimum*

1. Corolla dilating distally towards mouth 2
   Corolla not dilating towards mouth, ± parallel sided . . . . subgenus *Nautochilus*

2. Corolla dorsally gibbous at midpoint, posterior stamens usually appendiculate . . . . subgenus *Ocimum* . . . 3

   Corolla parallel-sided at midpoint, stamens never appendiculate . . . . subgenus *Gymnocimum*

3. Calyx throat with a hairy annulus . . . . section *Ocimum*
   Calyx throat glabrous 4

4. Bracts caducous, scar developing into a auxiliary nectary . . . . section *Hiantia* . . . 5

   Bracts persistent . . . . section *Gratissima* . . 8

5. Anthers elongate with parallel thecae, Old World . . . subsection *Hiantia* . . . 6
   Anthers rotund to reniform with divergent thecae, New World . . . . subsection *Nudicaulia*

6. Cymes 1-flowered . . . . series *Monobecium*
   Cymes 3-flowered 7

7. Sinus between lateral and median lobes of calyx lower lip . . series *Serpyllifolium*
   No sinus between lateral and median lobes of lower lip . . . . . . series *Hiantia*

8. Posterior lobe of calyx not membranous, nutlets ± spherical . . . .
   . . . . subsection *Gratissima*
   Posterior lobe of calyx membranous, nutlets obovate . . . .
   . . . . subsection *Erythrochlamys*

TAXONOMY AND ECONOMIC USE

Economically the most important taxon within *Ocimum* is section *Ocimum*. The most heavily used species are *O. basilicum*, *O. americanum* and their hybrid *O.* × *citriodorum*. These species are used for essential oil production and as pot herbs. *O. kilimandscharicum* is extensively grown in the tropics for camphor production and *O. forskolei*, like several other members of the group is used in traditional medicine as well as a pot herb (Demissew & Asfaw 1994). The relationships between these species are complex with many species hybridizing. This leads to the species being morphologically and chemically very variable (Paton 1996; Grayer *et al.*, 1996 a, b). A good account of the recognized forms of *O. basilicum* and relatives can be found in DeBaggio & Belsinger (1996). Attempts have been made in the past to distinguish between varieties of *O. basilicum*. However, as there is no clear morphological discontinuity between the varieties, the widespread use of varietal names obscures important information such as variation in chemistry and chromosome number. The cultivars and trade names used e.g., Swiss™ (Agrexco Agricultural Export Co. Ltd.) and cv. Cinnamon, are useful as they communicate several other characteristics such as scent, leaf shape and texture. However, there is a lack of standardisation in the descriptive terms used and the lack of verified voucher specimens for these trade forms makes it difficult assess the differences between them. A system of standardised descriptors is needed to allow easy communication and identification of the different forms of *O. basilicum*, *O.* × *citriodorum*, *O. americanum* and their close relatives. Such descriptors should include chemical information, in order to describe chemotypes, and chromosome numbers. This system of standardised descriptors is currently being used by IPGRI for the description of accessions of many crops, for example Cardamom (IPGRI 1994). Until such a system is in place for *Ocimum*, and particularly subsect. *Ocimum*, enabling accurate identification and unambiguous communication of different forms, full development of the economic and medicinal potential of this important group will remain an elusive goal.

O. gratissimum is another species of major economic importance with around 55 tonnes of essential oil produced annually from this species (Lawrence 1992) in addition to being used medicinally and as an insecticide (Githinji & Kokwaro 1993). None of the other species in sect. *Gratissima* seem to be used, but there may be potential for such use. Section *Hiantia* has few species of economic importance although studies of the essential oils of *O. nudicaule* have been madc (Morhy *et al.*, 1970) and *O. selloi* is cultivated. *O. obovatum* is used in traditional medicine in Ethiopia (Demissew & Asfaw 1994).

Ocimum lamiifolium is one of the most widely used medicinal herbs in Ethiopia (Demissew & Asfaw 1994), but little seems to be known about the uses of the remaining species of subgenus *Nautochilus*. Within subgenus *Gymnocimum*, *O. tenuiflorum* is widely used as a pot herb, medicinally and in essential oil production, *O. campechianum* Mill., is also used as a pot herb and medicinally.

The classification presented here will hopefully provide a framework for future exploration of medicinal and economic potential within *Ocimum*. Many species are as yet chemically unstudied. Knowledge of the relationships indicated here may lead the way to a more systematic study of potential uses of *Ocimum*.

KEY TO THE SPECIES OF *OCIMUM*

1. Bracts persistent, no bowl-like gland formed under cymes 2
   Bracts deciduous, bowl-like gland forming under cymes 30

2. Upper lip of the calyx markedly expanded in fruit, rotund, widest at middle, obscuring the rest of the calyx from above 3
   Upper lip of the calyx not markedly expanded in fruit, obovate, widest near tip, not obscuring the rest of the calyx from above 7

3. Posterior stamens sterile with small, poorly developed anthers . . . . **9. O. circinatum**
   Posterior stamens fertile 4

4. Posterior stamens appendiculate 5
   Posterior stamens inappendiculate . . . . **17. O. spectabile**

5. Plants tomentose, covered in dendroid hairs; upper lip of calyx thick textured 6
   Plants, pubescent with mostly simple and some dendroid hairs; upper lip of calyx membranous . . . . **16. O. fruticosum**

6. Leaves shorter than 15 mm long; nutlets glabrous . . . . **15. O. nummularium**
   Leaves longer than 15 mm long; nutlets pubescent . . . . . . **14. O. cufodontii**

7. Plants densely covered in dendroid hairs 8
   Plants, pubescent with simple, not dendroid, hairs 9

8. Inflorescence dense, calyces strongly reflexed against the inflorescence axis, touching those from adjacent verticils . . . . **12. O. spicatum**
   Inflorescence lax, verticils 5 – 15 mm apart; calyx horizontal or slightly downward pointing, not pressed against the inflorescence axis nor touching those from adjacent verticils . . . . **13. O. jamesii**

9. Throat of fruiting calyx closed, median lobes of lower lip pressed against upper lip, the lateral lobes level with or lower than the median lobes of the lower lip, interior of tube with hairs only underneath the upper lip; nutlets subspherical 10
   Throat of fruiting calyx open, upper and lower lips far apart, with teeth of lateral lobes held in between the upper lip and median lobes of the lower lip, interior of tube glabrous or with a dense ring of hairs at throat; nutlets ovoid . . 11

10. Leaves mostly more than 35 mm long, serrate, most petioles longer than 5 mm . . . .**10. O. gratissimum**
   Leaves less than 35 mm long, subentire or serrate in the distal portion, petioles less than 5 mm . . . .**11. O. natalense**

11. Calyx glabrous at throat; corolla-tube parallel-sided; posterior stamens with a clump of cilia near base; nutlets brown, not mucilaginous, or producing only a small amount, when wet 12
   Calyx with a dense ring of hairs at throat; corolla tube funnel-shaped; posterior stamen with a flattened or terete, glabrous or hairy outgrowth near base; nutlets black, producing copious mucilage when wet 21

12. Corolla less than 5 mm long 13
   Corolla longer than 5 mm 14

13. Fruiting calyx 6.5–10 mm long with the decurrent margin of the upper lip forming an erect conspicuous wing, posterior stamens glabrous . . . .**61. O. campechianum**
   Fruiting calyx 4–6 mm long with the decurrent margin of the upper lip inconspicuous, posterior stamens with a small tuft of hairs at base . . . .**64. O. tenuiflorum**

14. Corolla dilating towards throat, South American plants 15
   Corolla parallel-sided to throat, not dilating, African plants 16

15. Leaves sessile, petioles rarely 1 mm long . . . .**63. O. hassleri**
   Leaves petiolate usually longer than 3 mm, rarely 1–3 mm . . .**62. O. ovatum**

16. Lower lip of corolla 8–12 mm long . . . .**60. O. labiatum**
   Lower lip of the corolla less than 8 mm long 17

17. Corolla tube exceeding 20 mm long . . . .**57. O. tubiforme**
   Corolla tube 5–16 mm long 18

18. Cymes usually strictly 3-flowered 19
   Cymes with a variable number of flowers 2–7, but not uniformly 3-flowered 20

19. Flowering calyces 2–3 mm long; petioles of lower leaves as long as, or longer than the blade, blades mostly triangular . . . . **56. O. masaiense**
   Flowering calyces 4–5 mm long; petioles of lower leaves shorter than the blades, blades ovate . . . . **55. O. lamiifolium**

20. Leaf blade usually exceeding 40 × 20 mm; stamens exserted 4–6 mm from corolla throat . . . . **58. O. serratum**
    Leaf blade usually less than 40 × 20 mm; stamens exserted 7–8 mm from corolla throat . . . . **.59. O. pseudoserratum**

21. Bracts large, at least 6 mm wide, membranous, bright red or pink, obscuring young buds . . . . **.8. O. fischeri**
    Bracts small, less than 5 mm wide, ± thick-textured, greenish, rarely pinkish, rarely obscuring buds 22

22. Subshrub with lateral branches forming an angle of less than 30° with main branch; leaves mostly shorter than 20 × 10 mm . . . . **.3. O. minimum**
    Herb, subshrub or shrub with lateral branches forming an angle more than 30° with main branch; leaves various 23

23. Corolla more than 7 mm long; fruiting calyx 6 mm long or more 24
    Corolla up to 7 mm; fruiting calyx less than 6 mm long 25

24. Herb to 0.6 m high; stem glabrous or puberulent with minute hairs concentrated on two opposing faces of the stem; calyx pilose or pubescent . . . . **.4. O. basilicum**
    Shrub to 1 m high, pubescent with adpressed, retrorse hairs evenly dispersed around stem; calyx densely pubescent or hoary . . . . **.7. O. forskolei**

25. Leaves with apices mostly obtuse or rounded; stem with hairs patent 26
    Leaves with apices mostly acute or acuminate; stem with hairs patent or adpressed 27

26. Shrub to 2 m high; leaves mostly spreading; petioles more than 3 mm long . . . . **.1. O. kilimandscharicum**
    Herb to 0.3 m high; leaves mostly erect; petioles up to 3 mm long . . . . **2. O. kenyense**

27. Stem glabrous or minutely puberulent on two opposing faces . . . . **.4. O. basilicum**
    Stems pubescent with retrorse or patent hairs distributed equally around the stem 28

28. Fruiting calyx 6–8 mm long; stamens exserted 3–6 mm, shrubs . . . . **.7. O. forskolei**
    Fruiting calyx 4–6 mm long; stamens exserted 1–3 mm; herbs 29

29. Fruiting calyx 5–6 mm long; dried anthers usually more than 0.6 mm in diameter . . . . **.5. O. × citriodorum**
    Fruiting calyx 4–5 mm long; dried anthers usually around 0.5 mm in diameter . . . . **6. O. americanum**

30. Fruiting calyx bent downwards at throat forming a right angle with the tube; upper lip much expanded reniform . . . . **.18. O. transamazonicum**
    Fruiting calyx straight, upper lip not much expanded, obovate 31

31. Anthers elongate with parallel thecae; Old World plants 33
    Anthers orbicular to reniform with divergent thecae; New World plants . . 32

32. Leaves mainly in a basal rosette . . . . **20. O. nudicaule**
    Leaves regularly spaced up plant . . . . **19. O. selloi**

33. Cymes 1-flowered . . . . **21. O. irvinei**
    Cymes 3-flowered 34

34. Sinus between lateral and median lobes of calyx lower lip 35
    No sinus between lateral and median lobes of lower lip 39

35. Suffrutex with herbaceous stems arising from a woody rootstock 36
    Shrub or subshrub with woody stems 37

36. Leaves spreading or ascending, not clasping stem . . . . . **22. O. fimbriatum**
    Leaves erect clasping stem at base . . . . **23. O. amicorum**

37. Shrub 1–3 m tall; leaves 10–30 mm long; endemic to S. Ethiopia . . . . **24. O. ellenbeckii**
    Subshrub or shrub less than 1 m tall; leaves 5–20 mm long 38

38. Stamens exserted 10–14 mm long; pedicels rarely more than 3 mm long; endemic to the Arabian Peninsula and N Somalia . . . **25. O. serpyllifolium**
    Stamens exserted 6–9 mm; pedicels often more than 3 mm long; endemic to South Africa . . . . **26. O. burchellianum**

39. Posterior stamens inapendiculate; endemic to Oman . . . . **54.O. dhofarense**
    Posterior stamens appendiculate 40

40. Corolla 5 mm long or less 41
    Corolla more than 5 mm long 43

41. Annual herb; stamens exserted from corolla by only 2 mm . . . . **30. O. minutiflorum**
    Perennial woody herb or suffrutex; stamens exserted from corolla by at least 4 mm 42

42. Erect woody herb or suffrutex; leaves linear or narrowly obovate, always with fascicles of younger leaves in axils . . . . **27. O. angustifolium**
    Prostrate, carpet- or cushion-forming suffrutex; leaves ovate, obovate or elliptic, only occasionally with fascicles of younger leaves in axils . **39. O. mearnsii**

43. Indumentum of stem and leaves with dendroid or mixed dendroid and simple hairs 44
    Indumentum of stem and leaves with simple hairs only 48

44. Leaves ericoid, linear-lanceolate . . . . **36. O. monocotyloides**
    Leaves flat, ovate, obovate, ellipsoid 45

45. Leaves shorter than 30 mm long . . . . **34. O. metallorum**
    Leaves mostly longer than 30 mm long 46

46. Stem and leaves with simple and dendroid hairs; dendroid hairs branched only at base; lobes of upper lip of corolla fimbriate . . . . **48. O. obovatum subsp. crystallinum**
    Stem and leaves with dendroid hairs only; dendroid hairs branched along length; lobes of upper lip of corolla minutely dentate 47

47. Indumentum with densely branched dendroid, white hairs; shrub 0.5–2 m tall . . . . **32. O. albostellatum**
    Indumentum with laxly branched dendroid, often yellowish hairs; suffrutex 0.3 – 0.45 m tall . . . . **33. O. vanderystii**

48. Leaves linear to narrowly lanceolate, seemingly forming whorls through densely leaved axillary short shoots 49
    Leaves usually broader and not forming apparent whorls 51

49. Suffrutex with herbaceous stems arising from a woody rootstock; endemic to Zaire . . . . **35. O. ericoides**
    Shrub with woody stems; endemic to Ethiopia and Somalia 50

50. Corolla 12–15 mm long, stamens exserted 20–30 mm . . . . **28. O. formosum**
    Corolla 8–11 mm long; stamens exserted 10–18 mm . . . **29. O. verticillifolium**

51. Lateral lobe of the lower calyx-lip distally tomentose, not only on margin . . . . **40. O. centraliafricanum**
    Lateral tooth of the lower calyx-lip uniformly pubescent or ciliate at the margin only 52

52. Leaf venation largely parallel; endemic to Zaire 53
    Leaf venation largely divergent; throughout Tropical and South Africa . . . 54

53. Leaves sparsely pubescent, coriaceous; upper calyx lip obovate . . . . **42. O. mitwabense**
    Leaves hirsute, thin textured; upper calyx lip elliptic . . . **41. O. hirsutissimum**

54. Leaf blades folded and deflexed at least at apex, blades sometimes spreading to ascending at base 55
    Leaf blades spreading to erect, apex flat, rarely folded 61

55. Upper lip of calyx elliptic, apex acute . . . . **45. O. waterbergense**
    Upper lip of calyx obovate, apex rounded to mucronate 56

56. Leaves longer than 35 mm . . . . **53. O. pyramidatum**
    Leaves shorter than 35 mm 57

57. Leaves with revolute margin 58
    Leaves with flat margin 59

58. Apical bracts conspicuous; inflorescence condensed . . . **51. O. vandenbrandei**
    Apical bracts inconspicuous; inflorescence lax . . . . **52. O. urundense**

59. Leaves ascending, grey; stems virgate 60
    Leaves spreading or deflexed, rarely ascending, green; stems herbaceous, sometimes woody at base . . . . **50. O. decumbens**

60. Lobes of corolla upper lip fimbriate to dentate; endemic to Tanzania . . . . .49. **O. canescens**
    Lobes of corolla upper lip entire to sinuate . . . . .43. **O. dolomiticola**

61. Corolla tube dilated gradually; median lobes of corolla upper lip entire or denticulate . . . . .31. **O. filamentosum**
    Corolla tube dilated abruptly at throat; median lobes of corolla upper lip toothed or fimbriate, rarely sinuate 62

62. Upper calyx-tooth elliptic, apex acute 63
    Upper calyx tooth obovate, apex rounded or mucronate 64

63. Shrub 1–1.5 m tall, usually with a single stem arising from the ground . . . . .38. **O. vihpyense**
    Suffrutex, with several stems arising from a woody rootstock, less than 1 m tall . . . . .47. **O. dambicola**

64. Shrub 1–2.5 m tall; leaves less than 3 times longer than broad . .. . .37. **O. grandiflorum**
    Suffrutex or shrub up to 1.5 m tall; leaves more than 3 times longer than broad 65

65. Stems procumbent, mat-forming 66
    Stems mostly erect or ascending 67

66. Petioles 3–5 mm long, about half the length of the blade on mature leaves . . . . .46. **O. reclinatum**
    Petioles 0–2 mm long, much less than half the length of the blade on mature leaves . . . . 39. **O. mearnsii**

67. Petioles mostly longer than 3 mm . . . . .44. **O. coddii**
    Petioles less than 2 mm . . . . 48. **O. obovatum**

ENUMERATION OF SPECIES

Further information on the species can be found as follows: New World species (Epling 1936; Paton & Harley in prep.); African and Asian species, not subsect. *Hiantia* (Paton 1992, Paton 1996); African species subsect. *Hiantia* (Paton 1995a, Sebald 1988, 1989).

1. O. kilimandscharicum *Baker ex Gürke* in Pflanzenw. Ost-Afrikas. C: 349 (1895) (Figure 1.1d). Type: Tanzania, between Meru and Kilimanjaro, *Volkens* 756 (K, lectotype).
Distribution: Uganda, Kenya, Tanzania, introduced elsewhere in the tropics.
Habitat: Grassland, disturbed ground; 1200–1900 m.

2. O. kenyense *Ayobangira ex A.J.Paton* in Kew. Bull. 47: 422 (1992). Type: Kenya, Thika, *Bogdan* 4602 (holotype, K; isotype EA).

Distribution: Kenya and Tanzania.
Habitat: Wet places, seasonally waterlogged grassland; 1800–2000 m.

3. O. minimum *L.*, Sp. Pl.: 597 (1753). Type: Sri Lanka, Linnean Herbarium 749.6 (lectotype, LINN!).
Distribution: Asia, commonly cultivated.
Habitat: uncertain in the wild.

4. O. basilicum *L.*, Sp. Pl.: 597 (1753) (Figure 1.1c). Type: western Asia, Linnean Herbarium 749.5 (lectotype, LINN).
Distribution: Native in tropical Asia, perhaps NE Africa. Locally naturalized throughout tropical Africa, Asia and America. Cultivated in N Africa, Europe and SW. Asia.
Habitat: Cultivated, disturbed ground, ground prone to flooding, grassland.

5. Ocimum × citriodorum *Vis.* in Linnaea 15: Litteratur Bericht 102 (1841). Type: Without collector, cultivated at Padua (holotype, PAD).

Ozymum citratum Rumphius in Herbarium Amboinense 5: 266, t 93 figure 1.1. (1747).

Ocimum basilicum L var. *anisatum* Benth. in Labiat. Gen. Sp.: 4 (1832). Type: India, *Wallich* s.n. (lectotype, K).

O. dichotomum Hochst. ex Benth. in De Candolle, Prod. Syst. Nat. 12: 39 (1848). Type: Sudan, Kordofan, Arasch Kool, *Kotschy* 73 (lectotype, K).

O. americanum sensu Pushpangadan & Sobti *non* L. in Cytologia 47: 575–583 (1982).
Distribution: NE Africa, tropical Asia, widely cultivated.
Habitat: Cultivated, disturbed ground.

6. O. americanum *L.* in Cent. Pl. 1: 15 (1755). Type: America, Linnean Herbarium (749.9 LINN, lectotype).

i. var. **americanum**
O. canum Sims in Bot. Mag., *t.* 2452 (1823). Type: cultivated, seed from China, illustration in Curtis' Bot. Mag. *t.* 2452 (1853).
Distribution: Found widely in Tropical and southern Africa, China and India; naturalized in southern Europe and Australia and Tropical South America.
Habitat: Cultivated, disturbed ground, abandoned cultivation, grassland, often prone to flooding; sea-level to 1800 m.

ii. var. **pilosum** (*Willd.*) *A.J. Paton.* Type: probably cultivated in Berlin from seed of unknown origin, Willdenow Herbarium 11064 (B-Willd, lectotype, microfiche).
O. graveolens A.Br. in Flora 24: 265 (1840).
Distribution: Widespread in tropical Africa but absent in West Africa west of Cameroon, southern Africa, tropical Asia.
Habitat: Disturbed ground, damp areas.

7. O. forskolei *Benth.* Labiat. Gen. Sp. 1: 6 (1832). Type: N. Yemen, Hadie and Srd, Forsskål (BM, holotype; C, isotype).

O. menthiifolium Hochst. ex Benth. in DC., Prodr. 12: 34 (1848).

O. hadiense sensu E.A.Bruce in Kew Bull. 1935: 323 (1935) *non O. hadiense* Forssk.

O. staminosum Baker in Kew Bull. 1895: 224 (1895).

O. stirbeyi Schwein. & Volk., Pl. Ghika-Com.: 13 (1897).

Distribution: Arabia, Yemen, Egypt, Sudan, Somalia, Ethiopia, northern and eastern Kenya.
Habitat: Scrub, dry woods; 800m.

8. O. fischeri *Gürke* in Bot. Jahr. Syst. 19: 195 (1894). Type: Kenya, Kwale District, c.72 Km. from Mombasa on the Nairobi road, 20 Nov. 1962, *Greenway* 10855 (neotype, K).
Distribution: Endemic to E. Kenya and NE. Tanzania
Habitat: *Acacia-Commiphora* scrub, often on rocky ground; 300–500 m.

9. O. circinatum *A.J. Paton* Type: Ethiopia, Hararge Region, near Sullare, 22 Nov. 1953 *Popov* 1110 (holotype, K).
Distribution: Somalia: Hiiraan, Galguduud Provinces; Ethiopia: Hararge Region.
Habitat: Sandy and stony plains, bushland on orange sand over limestone; 130–420 m.

10. O. gratissimum L. Sp. Pl.: 1197 (1753). Type: cultivated in Uppsala, originally from India, Linnean Herbarium 749.2 (neotype, LINN).
a. subsp. **gratissimum**

i. var. **gratissimum**
O. urticifolium Roth, Catalecta Bot. 2: 52 (1800).
O. suave Willd. Enum. Pl. Hort. Berol.: 629 (1809).
O. viride Willd. Enum. Pl. Hort. Berol.: 629 (1809).
O. trichodon Gürke in Pflanzenw. Ost-Afrikas. C: 350 (1895).
Distribution: Widespread in the tropics from India to West Africa, as far south as Namibia and Natal, naturalised in Tropical South America.
Habitat: Submontane forest, dense bush, bush in wet savanna, burnt ground, disturbed land, coastal bush, lake shores; 0–1500 m.

ii. var. **macrophyllum** Briq. in Bull. Herb. Boiss. 2: 120 (1894). Type: Madagascar, Bourbon, *Boivin* s.n. (G, lectotype).
Distribution: Widespread in the tropics from India to West Africa; in tropical Africa found in Somalia, Kenya and Tanzania in the east and throughout W. Africa south to Angola; naturalized in Brazil and West Indies.
Habitat: Clearings in forest, coastal cliffs, disturbed ground; sea level–1000 m.

b. subsp. **iringense** (*Ayobangira*) *A.J. Paton* in Kew Bull. 47: 417 (1992). Type: Tanzania, Iringa district, Msembe-Mbagi track, *Greenway & Kanuri* 14000 (MO, holotype; K, isotype).
Distribution: Endemic to Shinyanga, Mpwapwa, Dodoma and Iringa districts of Tanzania.
Habitat: *Acacia-Commiphora-Adansonia* bushland, cultivated grazed areas, often on sandy soils; 700–1200 m.

11. O. natalense *Ayobangira ex A.J. Paton* in Kew Bull. 47: 417 (1992). Type: Natal, Mapelana forest, S. of St. Lucia estuary, *Cooper* 119 (holotype, PRE).
Distribution: Mozambique: Gaza, Inhaca Island; South Africa: Natal.
Habitat: Margins of dune forest; 0–200 m.

12. O. spicatum *Deflers* in Bull. Soc. Bot. Fr. 43: 226 (1896). Lectotype chosen here: N. Yemen, Jabal Masna'ah, *Deflers* 599 (K, lectotype)
Distribution: Yemen, Ethiopia, Kenya; Northern Frontier Province.
Habitat: Woodland on alluvial soil; 750–850 m.

13. O. jamesii *Sebald* in Stuttgarter Beitr. Naturk. Ser. A, 405: 3 (1987). Type: Somalia, Hahi, *James & Thrupp s.n.* (holotype, K).
O. tomentosum Oliver in Hook. Icon. Pl. 21: 1*t* 1539 (1886) nom. illeg., non *O.tomentosum* Lam. Encycl. 1: 387 (1785).
Habitat: *Acacia* and *Commiphora* scrub on limestone derived soil; 540–1200 m.
Distribution: Somalia: Woqooyi Galbeed, Saanag, Nugaal, Mudug, Bay, Bakool, Hiiraan, Togdheer Provinces; Ethiopia: Hararge Region.

14. O. cufodontii (*Lanza*) *A.J. Paton* in Grana 31: 162 (1992). Type: Ethiopia, Borana, Malca Guba on Daua Parma, *Cufodontis* 112 (holotype, FT).
Erythrochlamys cufodontii Lanza in Miss. Biol. Borana. Racc. 183, Figure 52 (1939);
Distribution: Ethiopia, Somalia, Kenya: Northern Frontier Province.
Habitat: Scrub on lava, scrub on limestone, Acacia bushland; 370–500 m.

15. O. nummularium (*S.Moore*) *A.J. Paton* in Grana 31: 162 (1992). Type: Somalia, Ahl Mountains, *Hildebrandt* 853 (BM, holotype).
Erythrochlamys nummularia (S.Moore) Hedge in Notes Roy. Bot. Gard. Edinburgh 35: 189 (1977).
Distribution: N. Somalia: Barr, Sanaag, Nugaal and Mudug Provinces.
Habitat: *Acacia* and *Commiphora* scrub, open limestone slopes and screes.

16. O. fruticosum (*Ryding*) *A.J. Paton* **comb. nov.** Type: Somalia, Hiiraan, 20 km E of Buulobarde (holotype, UPS; isotype K, MOG).
Basionym: *Erythrochlamys fruticosa* Ryding in Nord. J. Bot. 10: 633 (1991).
Distribution: Somalia, Galguduud & Hiiraan.
Habitat: *Acacia-Commiphora* woodland, on sandy soil overlying limestone or sandstone: 100–1000 m.

17. O. spectabile (*Gürke*) *A.J. Paton* **comb. nov.** Type: Kenya, Machakos District, between Ulu and Ukamba, *Fischer* 500 (holotype, B destroyed).
Basionym: *Erythrochlamys spectabilis Gürke* in Bot. Jahrb. Syst. 19: 223 (1894).
Distribution: Ethiopia, Somalia, Kenya, NE Tanzania.
Habitat: Dense bush, *Acacia*, *Commiphora*, *Combretum* bushland, lava slopes; 100–600 m.

18. O. transamazonicum *C.Pereira* in Bradea 1: 123 (1972). Type: Brazil, Goiás, Transamazonian highway, *Duarte* 13946 (holotype HB, isotype RFA).
Distribution. Brazil
Habitat. Grassland (cerrado).

19. O. selloi *Benth.*, Labiat. Gen. Sp.: 6 (1832). Type: Brazil, *Sello s.n.* (holotype, K).
O. carnosum (Spreng.) Link & Otto ex Benth., Labiat, Gen. Sp.:11 (1832). Type:

Brazil, *Sello s.n.*(holotype, B; isotype, K).
Distribution. Eastern S America.
Habitat. Scrub, wasteland, cultivated.

20. O. nudicaule *Benth.*, Labiat. Gen. Sp.: 14 (1832). Type: Brazil, *Sello* 4991 (lectotype, B; isotype, K).
Distribution. Brazil, Paraguay, Argentina.
Habitat. Grassland (cerrado).

21. O. irvinei *J.K. Morton* in J. Linn. Soc. Bot. 58: 266 (1962). Type: Ghana, S of the Adamsu junction of the Berekum to Sampa road. (holotype GC; isotype K).
Distribution: Ghana, Ivory Coast.
Habitat: Seasonally flooded grassland.

22. O. fimbriatum *Briq.* in Bot. Jahrb. Syst. 19: 161 (1894). Type: Angola, Malange, *Mechow* 165 (B, holotype; BR, isotype).
B. fimbriatum (Briq.) Sebald in Stuttgarter Beit. Naturk. Ser. A, 405: 10 (1987); Sebald in Stuttgarter Beitr. Naturk. A, 419: 42 (1988).

i. var. **fimbriatum** Type as for *O. fimbriatum*
B. fimbriatum (Briq.) Sebald var. *fimbriatum* in Stuttgarter Beitr. Naturk. A, 419: 45 (1988).
Distribution. Burundi, Zaire, Tanzania, Mozambique, Malawi, Zambia, Zimbabwe and Angola.
Habitat. Grassland, open *Brachystegia/Uapaca* woodland, edges of dambos (seasonally flooded areas) and areas prone to burning; 800–2160 m.

ii. var. **ctenodon** (*Gilli*) *A.J. Paton.* **comb. nov.** Type: Tanzania, Matengo highland WSW. of Songea, *Zerny* 117 (holotype, W).
Basionym: *Becium ctenodon* Gilli in Ann. Naturhist. Mus. Wien 77: 31 (1973).
B. fimbriatum (Briq.) Sebald var. *ctenodon* (*Gilli*) *Sebald* in Stuttgarter Beitr. Naturk. A, 405: 10 (1987).
Distribution: Zaire, Tanzania, Mozambique, Malawi and Zambia.
Habitat: *Brachystegia* woodland grassland prone to burning, often on sandy soil; 600–1800 m.

iii. var. **bequaertii** (*De Wild.*) *A.J. Paton* **comb. nov.** Type: Zaire, Bukama, *Bequaert* 111 (holotype, BR).
Basionym: *Becium bequaertii* De Wild., Contrib. Fl. Katanga: 179 (1921).
B. fimbriatum (Briq.) Sebald var. *bequaertii* (De Wild.) Sebald in Stuttgarter Beitr. Naturk. A, 419: 49 (1988).

iv. var. **angustilanceolatum** (*De Wild.*) *A.J. Paton* **comb. nov.** Type: Zaire, Shaba, plateau de Biano, Esschen Plateau, *Homblé* 866 (BR, holotype).
Basionym: *Ocimum angustilanceolatum* De Wild., Contrib. Fl. Katanga: 179 (1921); Ann. Soc. Sci. Bruxelles sèr B 41: 16 (1921).
B. fimbriatum (Briq.) Sebald var. *angustilanceolatum* (De Wild.) Sebald in Stuttgarter Beitr. Naturk. A, 405: 10 (1987).

Distribution: Zaire, Tanzania, Zambia.
Habitat: *Brachystegia* woodland, grassland on dark soil; 1290 m.

v. var. **microphyllum** (*Sebald*) *A.J. Paton.* **comb. nov.** Type: Zambia, Luwingu Distr., Chisinga Ranch, *Astle* 1006 (SRGH, holotype; K, isotype).
Basionym: *B. fimbriatum* (Briq.) Sebald var. *microphyllum* Sebald in Stuttgarter Beitr. Naturk. A 419: 51 (1988).
Distribution: Zaire and Zambia.
Habitat: Seasonally flooded grassland; 1380–1550 m.

23. O. amicorum *A.J. Paton* **nom. nov.** Type: Tanzania, Sumbawanga Distr.,Tatanda mission, *Bidgood, Mbago & Vollesen* 2381 (holotype, K).
Becium fastigiatum A.J. Paton in Kew. Bull. 50: 210 (1995).
Distribution: S Tanzania.
Habitat: Rocky hill with *Brachystegia* woodland; 1900 m.

24. O. ellenbeckii *Gürke* in Bot. Jarhb. Syst. 38: 172 (1906). Type: Ethiopia, Arussi-Galla Highland (holotype, B, destroyed).
Becium ellenbeckii (Gürke) Cufodontis, Enum. Pl. Aeth.: 849 (1963).
Distribution: S Ethiopia
Habitat: *Acacia*/*Commiphora* woodland; 1000–1200 m.

25. O. serpyllifolium *Forssk.*, Fl. Aegypt.-Arab.: 110 (1775). Type: Yemen, Jabal Khadra, Forsskål (holotype C- Forsskål).
Becium serpyllifolium (Forssk.) J.R.I. Wood in Kew Bull. 37: 602 (1983).
Distribution: Saudi Arabia, Yemen, N. Somalia.
Habitat: Rocky slopes; 1500–2000 m.

26. O. burchellianum *Benth., Labiat. Gen. Sp.*: 8 (1832). Type: South Africa, Middleburg Distr., *Burchell* 2812 (lectotype, K; isotype, PRE).
Becium burchellianum (Benth.) N.E.Br., Fl. Cap. 5: 232 (1910).
Distribution: South Africa, Eastern Cape.
Habitat: Scrub.

27. O. angustifolium *Benth.* in DC, Prodr. 12: 37 (1848). Type: South Africa, Transvaal, Magaliesberg, *Burke* s.n. (K, holotype).
Becium angustifolium (Benth.) N.E.Br. in Thistleton-Dyer, Fl. Cap. 5,1: 231 (1910).
Distribution: Kenya, Tanzania, Mozambique, Malawi, Zambia, Botswana, Zimbabwe, Angola & South Africa.
Habitat: Open *Brachystegia*/*Terminalia* woodland, grassland or dambos (seasonally flooded areas) often on sandy or black poorly drained soils and areas prone to burning; 850–2100 m.

28. O. formosum *Gürke* in Bot. Jahrb. Syst. 38: 173 (1906). Type: Ethiopia, Arusi-Galla highland, *Ellenbeck* 1953 (holotype, B, destroyed).
Becium formosum (Gürke) Chiov. ex Lanza in Miss. Biol. Borana 4: 182 (1939).

Distribution: Ethiopia, Bale province.
Habitat: *Acacia/Commiphora* woodland; 1700–1800 m.

29. O. verticillifolium *Baker* in Kew Bull. Misc. Inform. 1895: 224 (1895). Type: Somalia, Guldoo Hammed in M. Golis, *Cole* (syntype, K); Lort-Phillips (syntype, K)
Becium verticillifolium (Baker) Cufod., Enum. Pl. Aeth.: 851 (1963).
Distribution: Ethiopia and Somalia.
Habitat: *Juniperus procera* woodland, montane grassland; 1500–2000 m.

30. O. minutiflorum (*Sebald*) *A.J. Paton* **comb. nov.** Type: Burundi, Bubanza, Gihungwe, *Reekmans* 4281 (holotype, K; isotype, BR).
Basionym: *B. minutiflorum* Sebald in Stuttgarter Beitr. Naturk. A, 405: 11 (1987); Sebald in Stuttgarter Beitr. Naturk. A, 419: 59 (1988).
Distribution: Rwanda, Burundi, Zaire, Uganda, Tanzania, Zambia.
Habitat: Grassland, *Acacia-Commiphora* scrub or open mopane (*Colophospermum*) woodland on sandy soil; alt. 700–1710 m.

31. O. filamentosum *Forsskål*, Fl. Aegypt.-Arab.: 108 (1775). Type: Yemen, Mt Melhan, *Forsskål* 324 (holotype C, *Forsskål*).
Ocimum adscendens Willd., Sp. Pl. ed. 4, 3: 166 (1801).
Ocimum kenyanum Vatke in Linnaea 37: 315 (1871).
Becium filamentosum (Forssk.) Chiov. in Nuovo. Giorn. Bot. Ital., n.s. 26: 162 (1919).
Distribution: India, Saudi Arabia, Yemen, Sudan, Ethiopia, Uganda, Kenya, Tanzania, Mozambique, Botswana, Zimbabwe, Angola, Namibia and South Africa.
Habitat: Open woodland or grassland on basaltic, sandy or black cotton soils; 850–1950 m down to 20 m in coastal Kenya and Tanzania.

32. O. albostellatum (*Verdcourt*) *A.J. Paton* **comb. nov.** Type: Tanzania, Shinyanga, *Koritschoner* 2044 (holotype, EA; isotype, K).
Basionym: *Becium albostellatum* Verdcourt in Kew Bull. 1952: 364 (1952).
Distribution: Zaire, Tanzania, Zambia.
Habitat: *Brachystegia* or *Julbernardia* woodland, grassland prone to burning, often on sandy soil; 1050–1500 m.

33. O. vanderystii (*De Wild.*) *A.J. Paton* **comb. nov.** Type: Zaire, Kasai, Mukulu, *Vanderyst* 3273 (holotype, BR).
Basionym: *B. vanderystii* De Wild. in Ann. Soc. Sci. Bruxelles 41,2: 24 (1921).
Distribution: Zaire, Zambia & Angola.
Habitat: Open *Brachystegia* woodland on sandy soils, often on copper rich soils; 1500 m.

34. O. metallorum (*P.A. Duvign.*) *A.J. Paton* **comb. nov.** Type: Zaire, Katanga, Menga, *Duvigneaud & Timperman* 2098b (holotype, BRLU).
Basionym: *Becium metallorum* P.A. Duvign. in Trav. & Publ. Univ. Libre Lab. Bot. Syst. 26: 5 (1958); Lejeunia 21: 5 (1958).
B. grandiflorum Lam. var. *metallorum* (P.A. Duvign.) Sebald in Stuttgarter Beitr. Naturk. A, 437: 52 (1989).

Distribution: Zaire, Shaba.
Habitat: Grassland on copper rich soil; 1200–1600 m.

35. O. ericoides (*P.A. Duvign. & Plancke*) *A.J. Paton* **comb. nov.** Type: Zaire: Shaba, Chabara, *Duvigneaud* 4135B (holotype, BRLU).

Basionym: *Becium ericoides* P.A. Duvign. & Plancke in Bull. Soc. Roy. Bot. Belg. 96: 130 (1963)

B. grandiflorum Lam. var. *ericoides* (P.A. Duvign. & Plancke) Sebald in Stuttgarter Beitr. Naturk. A, 437: 55 (1989).

Distribution: Zaire, Shaba.
Habitat: Grassland on copper rich soil.

36. O. monocotyloides (*Ayobangira*) *A.J. Paton* **comb. nov.** Type: Zaire, Shaba, Mine de Kamatanda, *Plancke* 66/946 (holotype, BRLU).

Basionym: *Becium monocotyloides* Ayobangira in Etudes Rwandaise 1: 283 (1987).

Distribution: Zaire, Shaba.
Habitat: Copper bearing soil.

37. O. grandiflorum *Lam.*, Encycl. Meth. Bot. 1: 387 (1785). Type: Cultivated in Paris from seed collected by Bruce in Ethiopia (holotype, P-Lamark).

Becium grandiflorum (Lam.) Pic.Serm. in Webbia 7: 337 (1950).

B. bicolor Lindl. in Edwards's Bot. Reg. 28, Misc.: 42 (1842). Type: Cultivated in England from seed sent from Paris (CGE).

i. subsp. **densiflorum** *A.J. Paton* **comb. nov.** Type: Masai [Monuli] Distr., Merogai Gelai, 29 Sept. 1968, *Carmichael* 1505 (holotype K, isotype EA).

Becium grandiflorum (Lam.) Sebald subsp. *densiflorum* A.J. Paton in Kew Bull. 50: 217 (1995).

Distribution: N. Tanzania.
Habitat: Crevices in quarzite hills; 1685–2190 m.

ii. subsp. **turkanaense** (*Sebald*) *A.J. Paton.* **comb. nov.** Type: Kenya, Turkana, Murua Mts, *Newbould* 7063 (K, holotype; EA, isotype).

Becium grandiflorum (Lam.) Sebald subsp. *turkanaense* (Sebald) A.J. Paton in Kew Bull. 50: 219 (1995).

Distribution: N Kenya.
Habitat: Margins of *Juniperus* woodland; 1800–2250 m.

iii. subsp. **grandiflorum**

Distribution: Ethiopia.
Habitat: Steep rocky slopes; 2000–2500 m.

38. O. vihpyense A.J.Paton **nom. nov.** Type: Malawi, Mzimba Distr., Champira, *Pawek* 10492 (K, holotype; SRGH, isotype).

Becium frutescens (Sebald) A.J. Paton in Kew Bull. 50: 219 (1995).

Becium grandiflorum var. *frutescens* Sebald in Stuttgarter Beitr. Naturk. A, 437: 38 (1989). Type as above.

Distribution: Malawi & N Zambia.
Habitat: Among rocks in *Brachystegia* woodland or margins of evergreen forest.

39. O. mearnsii (*Ayob. ex Sebald*) *A.J. Paton* **comb. nov.** Type: Kenya, banks of the Undurango, Juja Farm near Nairobi, *Mearns* 55 (BM, holotype; C, isotype).

B. grandiflorum var. *mearnsii* Ayob. ex Sebald in Stuttgarter Beitr. Naturk. A, 437: 23 (1989).

B. mearnsii (Ayobangira ex Sebald) A.J. Paton in Kew Bull. 50: 222 (1995).
Distribution: Uganda, Kenya & N Tanzania
Habitat: Areas prone to flooding in open woodland or grassland, often on black soil or volcanic sand; 1524–1940 m.

40. O. centraliafricanum *R.E. Fries.* Wiss. Ergebn. Schwed. Rhod. Kongo-Exped 1: 286 (1916). Type: Zambia, near R. Kalungwisi, *R.E. Fries* 1166 (UPS, syntype); Zambia, Msisi near Mbala, *R.E. Fries* 1282 (UPS, syntype).

B. centraliafricanum (R.E. Fries) Sebald in Stuttgarter Beitr. Naturk. A, 405: 10 (1987).

Ocimum homblei De Wild., Contrib. Fl. Katanga: 180 (1921).

Becium homblei (De Wild.) Duvign. & Planke in Biol. Jaarb. 27: 239 (1959).
Distribution: Zaire, Tanzania , Zambia, Zimbabwe.
Habitat: Grassland on copper rich, often sandy soils, open *Brachystegia* woodland, dambos (seasonally flooded areas); 450–1500 m.

41. O. hirsutissimum (*P.A. Duvign.*) *A.J. Paton* **comb. nov.** Type: Zaire, Mitwaba, *Duvigneaud & Timperman* 2699 (holotype, BRLU).

Basionym: *Becium hirsutissimum* P.A. Duvign. in Bull. Soc. Roy. Bot. Belg. 90: 228 (1958).
Distribution: Zaire, around Mitwaba.
Habitat: Grassland.

42. O. mitwabense (*Ayobangira*) *A.J. Paton* **comb. nov.** Type: Zaire, Mitwaba, *Duvigneaud & Timperman* 2691 (holotype, BRLU).

Basionym: *Becium mitwabense* Ayobangira in Etudes Rwandaise 1: 277 (1987).
Distribution: Zaire, Mitwaba.
Habitat: Grassland.

43. O. dolomiticola *A.J. Paton* **nom. nov.**

Becium citriodorum Williamson & Balkwill in Kew Bull 50: 744 (1995). Type: South Africa, Transvaal, *Gesell* 21 (holotype, J; isotypes B, E, K).
Distribution: South Africa, Transvaal.
Habitat: Grassland on rocky hillsides with clay soils derived from dolomite.

44. O. waterbergense (*Williamson & Balkwill*) *A.J. Paton* **comb. nov.** Type: South Africa, Transvaal, *Balkwill & Balkwill* 4355 (holotype J; isotypes, E, K, LISC, M, MO).

Basionym: *Becium waterbergensis* Williamson & Balkwill in Kew Bull. 50: 747 (1995).
Distribution: South Africa, Transvaal.
Habitat: Grassland on rocky hillsides on sandstone derived soil.

45. O. coddii (*Williamson & Balkwill*) *A.J. Paton* **comb. nov.** Type: South Africa, Transvaal, *Balkwill & Cadman* 3783 (holotype, J; isotypes, E, K, MO, PRE).

Basionym: Becium coddii Williamson & Balkwill in Kew Bull. 50: 746 (1995).
Distribution: South Africa, Transvaal.
Habitat: Grassland on rocky hillsides with clay soils derived from dolomite or dry stream banks which have their origin in dolomitic areas.

46. O. reclinatum *Williamson & Balkwill* **comb. nov.** Type: South Africa, Kwazulu-Natal, Ngwavuma Distr., I.D.C. Rice Project, 2 km from turnoff to Phelendaba on Mbazwana road, *Germishuisen* 3612 (holotype, PRE).

Basionym: *Becium reclinatum* Williamson & Balkwill in Kew Bull. 50: 748 (1995).
Distribution: S Mozambique & South Africa, Kwazulu Natal.
Habitat: Coastal or riverine bush; 200 m.

47. O. dambicola *A.J. Paton* **nom. nov.** Type: Tanzania, lower plateau N. of L. Nyassa, *Thomson* s.n. (holotype, K).

O. punctatum Baker in Oliver, Fl. Trop. Afr. 5: 345 (1900). non. *O. punctatum L.* f.
Becium obovatum subsp. *punctatum* (Baker) A.J. Paton in Kew Bull. 50: 232 (1995).
Becium grandiflorum var. *galpinii* (Gürke) Sebald *pro parte* in Stuttgarter Beitr. Naturk. A, 437: 44 (1989).
Distribution: Tanzania, Malawi, Zambia.
Habitat: Grassland prone to burning, often on wet or peaty soil; 1800–2400 m.

48. O. obovatum E. Mey. ex Benth in E. Mey., Comm. pl. Afr. Austr.: 226 (1838). Type: South Africa, Natal, Umzimkulu, *Drège* 4769 (holotype, K).

Becium obovatum (E. Mey. ex Benth.) N.E.Br. in Thistleton-Dyer, Fl. Cap. 5,1: 230 (1910).

B. grandiflorum var. *obovatum* (E.Mey. ex Benth) Sebald in Stuttgarter Beitr. Naturk. A, 437: 29 (1989).

i. subsp. **obovatum**
a. var. **obovatum**
Distribution: Widespread in tropical Africa, in South Africa and in Madagascar.
Habitat: Grassland, open *Acacia* or *Brachystegia* woodland, often in areas prone to burning or seasonly flooded areas; 100–2100 m alt.

b. var. **galpinii** (*Gürke*) *A.J. Paton* **comb. nov.** Type: South Africa, Transvaal, Barberton, Saddleback ridge, *Galpin* 413 (isotype, PRE).
Basionym: *Ocimum galpinii* Gürke in Bot. Jahrb. Syst. 26: 78 (1888).
Becium obovatum var. *galpinii* (*Gürke*) *N.E.Br.* in Thistleton-Dyer, Fl. Cap. 5,1: 231 (1910).
Becium grandiflorum var. *galpinii* (Gürke) Sebald *pro parte* in Stuttgarter Beitr. Naturk. A, 437: 44 (1989). See also *B. obovatum* subsp. *punctatum.*
Distribution: Tropical and Southern Africa.
Habitat: Grassland, open *Brachystegia*, *Uacapa*, *Protea* woodland; 1500–2450 m.

ii. subsp. **cordatum** (*A.J. Paton*) *A.J. Paton* **comb. nov.** Type: Tanzania, top of Chunya escarpment, *Richards* 20743 (K, holotypus).
Basionym: *Becium obovatum* subsp. *cordatum* A.J. Paton in Kew Bull. 50: 234 (1995).

Distribution: Tanzania (T 7).
Habitat: Grassland prone to burning; 1400–2400m.

iii. subsp. **crystallinum** (*A.J. Paton*) *A.J. Paton* **comb. nov.** Type: Nyika, Nat. Park, Sawi Valley, 12 Nov. 1972, *Synge* WC 451 (K, holotypus).
Basionym: *Becium obovatum* subsp. *crystallinum* A.J.Paton in Kew Bull. 50: 234 (1995).
Distribution: S Tanzania and N Malawi.
Habitat: Dry grassland and *Brachystegia* woodland; 1980–2300 m.

49. O. canescens *A.J. Paton* **nom. nov.** in Kew Bull. 50: 235 (1995). Type: Tanzania, Mpitin, South Serengeti, *Newbould* 6425 (holotype, K; isotype, EA).
Becium virgatum A.J. Paton in Kew Bull. 50: 235 (1995).
Distribution: Tanzania.
Habitat: Overgrazed grassland prone to burning, limestone; 1500–1800 m.

50. O. decumbens *Gürke* in Bot. Jahrb. Syst. 30: 400 (1901). Type: Tanzania, Mbeya/ Chunya Distr., Unyiha, *Goetze* 1458 (holotype, B, destroyed; isotypes, BM, E).
Ocimum capitatum Baker in Oliver, Fl. Trop. Afr. 5: 345 (1900) nom. illeg., non *O. capitatum* Roth (=*Acrocephalus*).
B. capitatum Agnew, Upland Kenya Wild Flowers: 646 (1974).
B. grandiflorum (Lam.) Pic.Serm. var. *decumbens* (Gürke) Sebald in Stuttgarter Beitr. Naturk. A, 437: 25 (1989).
B. grandiflorum (Lam.) Pic.Serm. var. *capitatum* (Agnew) Sebald in Stuttgarter Beitr. Naturk. A, 437: 20 (1989).
Becium decumbens (Gürke) A.J. Paton in Kew Bull. 50: 237 (1995).
Distribution: Zaire, Uganda, Kenya, Tanzania, Zambia, Malawi and Angola.
Habitat: Grassland prone to burning sometimes in areas prone to flooding, also recorded from edge of *Juniperus* forest; 1130–3100 m

51. O. vandenbrandei (*Ayobangira*) *A.J. Paton* **comb. nov.** Type: Zaire, Shaba, 20 km S of Pepa, *Duvigneaud* 3758 (holotype, BRLU).
Basionym: *Becium vandenbrandei* P.A. Duvign. & Plancke ex Ayobangira in Etudes Rwandaise 1: 284 (1987).
Distribution. Zaire, Marungu Mts.
Habitat. Grassland.

52. O. urundense *Robyns & Lebrun* in Rev. Zool. Bot. Africaines 16: 371 (1928). Type: Burundi, Kitega, *Elskens* 234 (BR, holotype)
Basionym: *Becium urundense* (Robyns & Lebrun) A.J.Paton in Kew Bull. 50: 239 (1995).
B. grandiflorum var. *urundense* (Robyns & Lebrun) Sebald in Stuttgarter Beitr. Naturk. A, 437: 24 (1989).
Distribution: Tanzania and Burundi.
Habitat: Grassland prone to burning; 1500 m.

53. O. pyramidatum (*A.J. Paton*) *A.J. Paton* **comb. nov.** Type: Tanzania, Mpanda Distr., Silkcub highlands, *Richards* 7127 (holotype, K).

Basionym: *Becium pyramidatum* A.J. Paton in Kew Bull. 50: 239 (1995).
Distribution: W Tanzania.
Habitat: Grassland or open *Brachystegia-Uapaca* woodland; 1050–1590 m.

54. O. dhofarense (*Sebald*) *A.J. Paton* **comb. nov.** Type: Oman, Dhofar Mts, *Bent* 152 (holotype, K).
Basionym: *Becium dhofarense* Sebald in Stuttgarter Beitr. Naturk. A, 405: 7 (1987).
Distribution: Oman, Dhofar.
Habitat: Escarpment woodland; 200–700 m.

55. O. lamiifolium *Hochst. ex Benth.* in De Candolle, Prodr. 12: 37 (1848). Type: Ethiopia, Scholada, *Schimper* 147 (lectotype, K).
Distribution: Cameroon, Zaire, Rwanda, Ethiopia, Kenya, Uganda, Tanzania, Malawi, Zambia.
Habitat: Scrubby forest, clearings, wasteland, secondary bush, forest edges; 1000–2100.

56. O. masaiense *Ayobangira ex Paton* in Kew Bull. 47: 431 (1992). Type: Kenya, Magadi road, *Glover & Samuel* 2917 (EA, holotype, isotype K).
Distribution: Kenya (Ngong Hills).
Habitat: *Acacia* scrub, between boulders on lava slopes.

57. O. tubiforme (*R.Good*) *A.J. Paton* **comb. nov.** Type: South Africa, Transvaal, Vaalhoek, Rogers 25104 (holotype, BM; isotype, PRE).
Basionym: *Orthosiphon tubiformis* R.Good in J. Bot., London 63: 173 (1925).
Distribution: South Africa (Transvaal).
Habitat: Wooded, stony slopes.

58. O. serratum (*Schlechter*) *A.J. Paton* **comb. nov.** Type: South Africa, Transvaal, Baberton, *Galpin* 499 (holotype, K; isotype PRE).
Basionym: *Orthosiphon serratus* Schlechter in J. Bot., London 35: 431 (1897).
Distribution: Swaziland, South Africa (Transvaal and KwaZulu Natal).
Habitat: Grassland prone to burning on stony hillsides.

59. O. pseudoserratum (*Ashby*) *A.J. Paton* **comb. nov.** Type: South Africa, Transvaal, Moorddrift, *Leedertz* 2243 (holotype, BM; isotype PRE).
Basionym: *Orthosiphon pseudoserratus* Ashby in J. Bot., London 76: 8 (1938).
Distribution: South Africa (Transvaal).
Habitat: Rocky, wooded slopes.

60. O. labiatum (*N.E.Br.*) *A.J. Paton* **comb. nov.** Type: South Africa, Transvaal, Woodbush, *Schlechter* 4434 (holotype, K; isotype, PRE).
Basionym: *Orthosiphon labiatus* N.E.Br. in Thistleton-Dyer, ed. Flora Capensis 5,1: 245 (1910).
Nautochilus labiatus (N.E.Br.) Bremek. in Ann. Transv. Mus. 15: 253 (1933).
Orthosiphon amabilis (Bremek.) Codd in Bothalia 8: 157 (1964).

Distribution: Zimbabwe, Swaziland, South Africa (Transvaal, KwaZulu Natal).
Habitat: Dry rocky, wooded hillsides and watercourses.

61. O. campechianum *Mill.*, Gard. Dict. ed 8, 5 (1768). Type: Central America (holotype, BM)

O. micranthum Wild., Enum. Hort. Berol. 630 (1809).

Distribution: Widespread in Tropical and Central America and West Indies.
Habitat: Disturbed ground.

62. O. ovatum *Benth.*, Labiat. Gen. Sp.: 13 (1832). Type: Brazil, *Sello* B1472 (lectotype, B; isotype K).

O. neurophyllum Briq. in Bull. Herb. Boiss. sér II 7: 263 (1907).

O. tweedianum Benth. in DC Prodr. 12: 38 (1948)

O. procurrens Epling in Fedde Rep. Beih. 85:182 (1936).

Distribution: S Brazil, Paraguay, Argentina.
Habitat: Grassland (cerrado).

63. O. hassleri *Briq.* in Bull. Herb. Boiss. sér. 2, 637 (1907) (Figure 1.1b). Type: Paraguay, *Hassler* 4380 (lectotype G; isotype K).
Distribution: Paraguay.
Habitat: Grassland (cerrado).

64. O. tenuiflorum L., Sp. Pl.: 597 (1753). Type: cultivated at Uppsala, Linnean Herbarium 749.13 (lectotype, LINN).

O. sanctum L, Mant. Pl: 85 (1767).

Distribution: Widespread in Tropical Asia, widely cultivated.
Habitat: Disturbed ground, cultivated.

Species excluded

65. O. tashiroi *Hayata*, Ic. Pl. Formosana 9: 86 (1920). Type: Taiwan, Holisha, *Tashiro, s.n. ann.* March 1896. This name, recorded from Taiwan is *Basilicum polystachyon* (L.) Moench (Paton & Cafferty 1998).*

* New reference

REFERENCES

Bentham, G. (1832). *Ocimum.* In *Labiatarum Genera and Species*: London, Ridgeway, 1–19.

Bentham, G. (1848). *Ocimum.* In Candolle, A.P. de, ed., *Prodromus Systematis Naturalis,* Paris, **12**, 31–44.

Briquet, J. (1897). *Ocimum.* In Engler, A. & Prantle, K.A.E. *Die Natürlichen Pflanzenfamilien* **4 & 3a**, 369–372.

Codd, L.E. (1964). The South African species of *Orthosiphon. Bothalia* **8**, 146–162.

Codd, L.E. (1985). Labiatae in: Liestner, O.A. ed., *Flora of Southern Africa.* South Africa, Pretoria, Botanical Research Institute, **28, 4.**

Darrah, H. (1980). *The cultivated Basils.* Thomas Buckeye Printing Company, Independence, Missouri.

DeBaggio, T. and Belsinger, S. (1996). *Basil, the herb lover's guide.* Interweave Press, Loveland, Colorado, USA.

Demissew, S. and Asfaw, N. Some useful indigenous labiates from Ethiopia. *Lamiales Newsletter* **3**, 5–6.

Epling, C. (1936). *Ocimum* in Synopsis of South American Labiatae. *Fedde Rep. Beih.* **85**, 180–184.

Githinji, C.W. and Kokwaro, J.O. (1993). Ethnomedicinal study of major species in the family Labiatae from Kenya. *J. Ethnopharmacology* **39**, 197–203.

Grayer, R.J., Kite, G.C., Goldstone, F.J., Bryan, S.E., Paton, A. and Putievsky, E. (1996 a). Infraspecific taxonomy and essential oil chemotypes in sweet basil, *Ocimum basilicum. Phytochemistry* **43**, 1033–1039.

Grayer, R.J., Bryan, S.E., Veitch, N.C. Goldstone, F.J., Paton, A. and Wollenweber, E. (1996 b). External flavones in sweet basil, *Ocimum basilicum*, and related taxa. *Phytochemistry* **43**, 1041–1047.

Greuter, W. *et al.* (1994). *International Code of Botanical Nomenclature.* Koeltz Scientific Books, Königsteir, Germany.

Gürke, M. (1894). *Erythrochlamys* in *Bot. Jahrb. Syst.* **19**, 222–223.

Harley, M.M., Paton, A., Harley, R.M. and Cade, P.G. (1992). Pollen morphological studies in tribe Ocimeae (Nepetoideae: Labiatae): 1. *Ocimum* L. *Grana* **31**, 161–176.

Hedge, I.C. (1992). A global survey of the biogeography of the Labiatae. In: Harley, R.M. and Reynolds, T. eds. *Advances in Labiate Science*: Royal Botanic Gardens, Kew. UK. 7–18.

Hedge, I.C. and Miller, A.G. (1977). New and interesting taxa from NE. Tropical Africa. *Notes Roy. Bot. Gard. Edinburgh* **35**, 179–193.

IPGRI (1994). Descriptors for Cardamom (*Elettaria cardamomum* Maton). International Plant Genetic Resources Institute. Rome.

Lawrence, B.M. (1992). Chemical components of Labiatae oils and their exploitation. In: Harley, R.M. and Reynolds, T. eds. *Advances in Labiate Science*: Royal Botanic Gardens, Kew. UK. 399–436.

Lindley, J. (1842) *Becium bicolor* in *Edward's Bot. Reg.* **28**, 42–43.

Linnaeus, C. (1753). *Ocimum.* In *Species Plantarum*, ed. 1. Holmiae [Stockholm], Laurentii Salvii. 597–598.

Morhy, L., Gomes, J. and Laboriau, J.G. (1970). *Ocimum nudicaule* Benth., A new source of methylchavicol. *An. Acad. Bras. Cienc.* **42**, 147–158.

Paton, A. (1992). A synopsis of *Ocimum* L. (Labiatae) in Africa. *Kew Bull.* **47**, 405–437.

Paton, A. (1994). A revision of the genus *Endostemon* (Labiatae). *Kew Bull.* **49**, 673–716.

Paton, A. (1995). The genus *Becium* in East Africa. *Kew Bull.* **50**, 199–242.

Paton, A. (1995b). A new species and new combinations in *Orthosiphon* and *Fuerstia* (Labiatae). *Kew. Bull.* **50**, 147–150.

Paton, A. (1997). Classification and species of *Platostoma* and *Haumaniastrum* (Labiatae). *Kew Bull.* **52**, 257–291.

Paton, A. and Cafferty, S. (1998). The identity of *Ocimum tashiroi* Hayata (Labiatae). *Kew Bulletin* **53**: 466.

Paton, A. and Putievsky (1996). Taxonomic problems and cytotaxonomic relationships between and within varieties of *O. basilicum* and related species (Labiatae). *Kew. Bull.* **51**, 509–524.

Pereira, C. (1972). Contribuição ao conhecimento da familia "Labiatae". –1. *Bradea* **1**, 123–128.

Pushpangadan, P. (1974). *Studies on reproduction and hybridization of Ocimum species with view to improving their quality.* Ph.D. thesis, Aligarh Muslim University, Aligarh, India.

Pushpangadan, P. and Bradu, B.L. (1995). Basil. In Chadha, K.L. & Rajendra Gupta, eds, *Advances in Horticulture Vol.* **11**– *Medicinal and Aromatic Plants.* Malhotra Publishing House, New Delhi.

Ramamoorthy, T.P. (1986). A revision of *Catoferia* (Labiatae). *Kew Bull.* **41**, 299–305.

Ryding, O. (1991). Notes on the genus *Erythrochlamys* (Lamiaceae). *Nord. J. Bot.* **10**, 633–635.

Ryding (1992). Pericarp structure and phylogeny within Lamiaceae subfamily *Nepetoideae* tribe *Ocimeae. Nord. J. Bot.* **12**, 273–298.

Sebald, O. (1988). Die Gattung *Becium* Lindley (Lamiaceae) in Afrika und auf der Arabischen Halbinsel. Tiel 1. *Stuttgarter Beitr. Naturk. A*, **419**.
Sebald, O. (1989). Die Gattung *Becium* Lindley (Lamiaceae) in Afrika und auf der Arabischen Halbinsel. Tiel 2. *Stuttgarter Beitr. Naturk. A*, **437**.
Sobti, S.N. and Pushpangadan, P. (1979). Cytotaxonomical studies in the genus *Ocimum*. In Bir, S.S., ed., Recent Researches in Plant Sciences: 373–377. Kalyani Publishers, New Delhi.
Swofford, D.L. (1993). *Phylogenetic analysis using parsimony (PAUP), version* **3.1**. Illinois Natural History Survey. Champaign.
Vogel, S. (1998). Remarkable nectaries: structure, ecology, organophyletic perspectives IV. Miscellaneous cases. *Flora* **193**: 225–248.

APPENDIX 1

Data matrix used in parsimony analysis.

O.americanum	00000	10000	20000	10121	11001	10000	01000	0
B.grandiflorum	11000	00000	12010	10021	11101	10000	01001	0
O.selloi	11000	00000	00000	00111	11001	10000	10000	0
O.tenuiflorum	00000	00000	00000	00100	1?001	10000	10000	0
B.irvinei	11010	00000	00000	10121	11101	10000	01000	0
B.dhofarense	11000	00000	12010	10011	11101	10000	01001	0
B.fimbriatum	11000	00000	02010	10021	11101	10000	01001	0
O.lamiifolium	00000	00000	00000	11001	11001	10000	00000	0
Orth.tubiformis	00020	00000	00000	01000	11001	10000	00100	0
Orth.labiatus	00020	00000	00000	11001	11001	10000	00100	0
O.gratissimum	00000	00000	01100	10121	11001	10100	10000	0
O.jamesii	00100	00000	01100	10121	11001	10110	10000	0
O.cufodontis	00100	00100	01100	10121	11001	10110	10000	0
E.spectabilis	A0100	00110	00100	10100	11001	10000	00100	0
E.fruticosus	A0120	00110	00100	10121	11001	10000	00A00	0
O.transamazonicum	11000	00110	00001	10121	11001	10000	10000	0
O.nudicaule	11000	00000	00000	10121	11001	10000	10000	0
O.ovatum	00000	00000	00000	00100	01011	10000	10000	0
O.campechianum	00000	00000	00000	00100	0?011	10000	10000	0
C.chiapensis	00001	01001	31010	00100	00000	00000	00011	1
C.capitata	00001	01001	20010	00000	00000	00000	00011	1
Orth.aristatus	00000	00001	20000	00000	00000	00000	00000	0
Orth. subgen. Orth.	000A0	00000	00000	00000	A0000	00000	00000	0
Syncolostemon	100C0	01000	20000	00000	A1001	01001	00011	1
Hemizigia	100B0	00000	00000	00000	A100A	01001	00011	1
O.circinatum	00000	10100	20000	10121	11001	10000	A1000	0

A= 0&1 B= 0&1&2 C= 1&2

Taxon Key.

O. = *Ocimum*; B. = *Becium*; E. = *Erythrochlamys*; S. = *Syncolostemon*; H. = *Hemizygia*; C. = *Catoferia*; Orth. = *Orthosiphon*.

2. PRODUCTION SYSTEMS OF SWEET BASIL

ELI PUTIEVSKY[1] and BERTALAN GALAMBOSI[2]

[1]*Agricultural Research Organization, Newe Ya'ar Research Centre, P.O Box 90000, Haifa 31900, Israel*

[2]*Agricultural Research Centre of Finland, Karila Research Station for Ecological Agriculture, Karilantie 2 A, FIN-50600 Mikkeli, Finland*

MAIN ELEMENTS OF SWEET BASIL CULTIVATION TECHNIQUES

During the long cultivation history of basil, numerous experiments have been carried out to specify the elements of basil cultivation techniques suitable for different climatic regions (Darrah 1972). For example, in the Horticultural Abstract Literature review during 1973–1993 more than 105 publications deal with agronomical matters. As a result of the research and the practical efforts, there is a great variability in the cultivation methods of basil around the world.

Ecology

Sweet basil is a tender herbaceous annual plant, which originates from tropical and warm areas, such as India, Africa and southern Asia. It is naturalized almost all over the world. Basil is reported to tolerate very variable ecological circumstances. It grows in the cool moist and tropical rain forest zones in annual temperatures between 6 and 24°C and receiving 500–8000 mm annual precipitation (Duke and Hurst 1975).

Although basil is cultivated in different climatic and ecological conditions, the most favourable conditions are found in countries with a warm climate. Warmth, light and moisture are the basic ecological requirements for basil cultivation. It is commonly known that basil is rather susceptible to frost. There are numerous research results reported from countries with a temperate climate. These results indirectly proved the warmth requiring characteristics of basil (Hälvä 1987b, Nykänen 1989, Sorensen and Henriksen 1992).

Light

Basil develops best under long days in sunny conditions. In a controlled experiment the average plant weight of *Ocimum basilicum* var. *citriodora* ranged between 90–98 g/pot at 15–21 hours of light. The flowering stage appeared rapidly when the plants were exposed to 18 hours of light, and the greatest yield (102 g/pot) was obtained under 24 hours of light (Skrubis and Markakis 1975).

Temperature

The optimum day/night temperature for seed germination was determined to be 24–27/19–22°C in laboratory conditions. At these temperatures over 80% seed germination was reached after four days. (Putievsky 1983) (Figure 2.1).

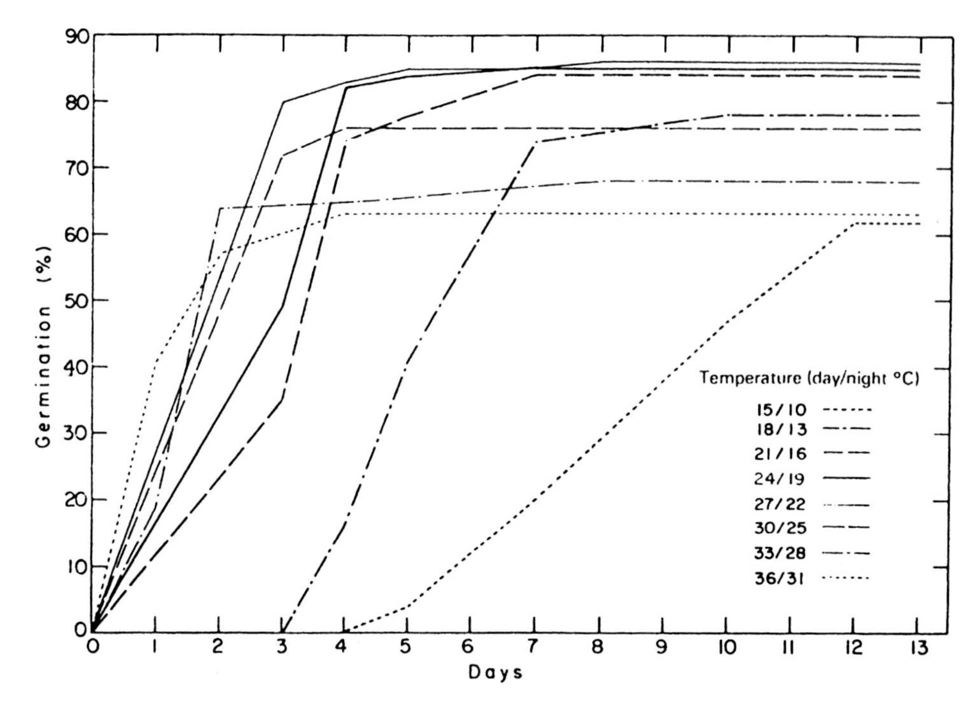

Figure 2.1 Effect of temperature on germination of sweet basil (Putievsky 1983).

In a growth chamber experiment the fastest growth rate was measured at a temperature of 27°C, and seven harvests were obtained. At a temperature of 32°C the number of harvests was only four and at 21°C five harvests were obtained (Pogany *et al.*, 1968).

In an Italian study sweet basil (*O. basilicum* cv. Grande verde) was grown in plain and mountain locations at 380 m and 1050 m altitudes, respectively. In the warmer plain conditions the fresh weight was two to three times higher than in the cooler mountain conditions. The essential oil content of basil grown in the mountains was lower (0.78–1.73%) than in the plain (1.25–2.40%) (Menghini *et al.*, 1984).

Sweet basil grown in Scotland gave a fresh yield of 4.6–6.8 kg/m^2 and a dry yield of 0.5–1.0 kg/m^2. These yields are comparable with yields obtained in southern countries only in greenhouse at a temperature of 18°C (Hay *et al.*, 1988).

In South Finland (60° and 61° northern latitude) Agryl P17 fiber cloth mulches increased the fresh basil yield more than three times. Further north, at 69° latitude, basil practically did not give any yield (Hälvä 1987b).

In a Danish study the additional warmth obtained by covering the soil increased the plant establishment and the fresh yield. A 0.05 mm thick, transparent plastic sheet with 500 holes/m^2 was placed on the soil immediately after sowing. Covering of the field increased the fresh yield by 35–70%, and also increased the number of surviving plants by more than 100% (Sorensen and Henriksen 1992).

In another Finnish experiment sweet basil of Hungarian origin was grown in the field and in a plastic house in Puumala, South Finland (61°, 40'). The warm indoor conditions increased the fresh yield, and doubled the volatile oil content as well (Nykänen 1986). The eugenol content of fresh basil in the open field was 67–98 mg/kg and in the plastic house 246–259 mg/kg (Figure 2.2).

When a methyl cinnamate chemotype of sweet basil was grown in the field and in greenhouse conditions at day/night temperatures of 26–30/18–21°C the essential oil content and the proportions of *cis*-methyl cinnamate, linalool and 1,8-cineole were significantly higher than in field grown plants. However, the content of *trans*-methyl cinnamate was higher in plants grown indoors (Morales *et al.*, 1993).

Soil

Basil is cultivated in the field in different types of soil with a pH of 4.3 to 8.2. Basil likes moderately fertile or humus-rich, well drained loamy or sandy-loam soils. The best soils are those which are in good physical condition and have a good water holding capacity. Waterlogged lands should invariably be avoided. In Egypt, reclaimed soil can be used for basil production by improving the soil quality with 30–40 t/ha organic matter.

In greenhouses basil is grown in beds or pots in different substrates, such as peat, different ready made mixtures or artificial soils (Grodan). Mainly for the production of fresh basil the plants are cultivated in beds or in different hydroponic systems such as aeroponics and nutrient film technique (Davide and Steward 1986) or using

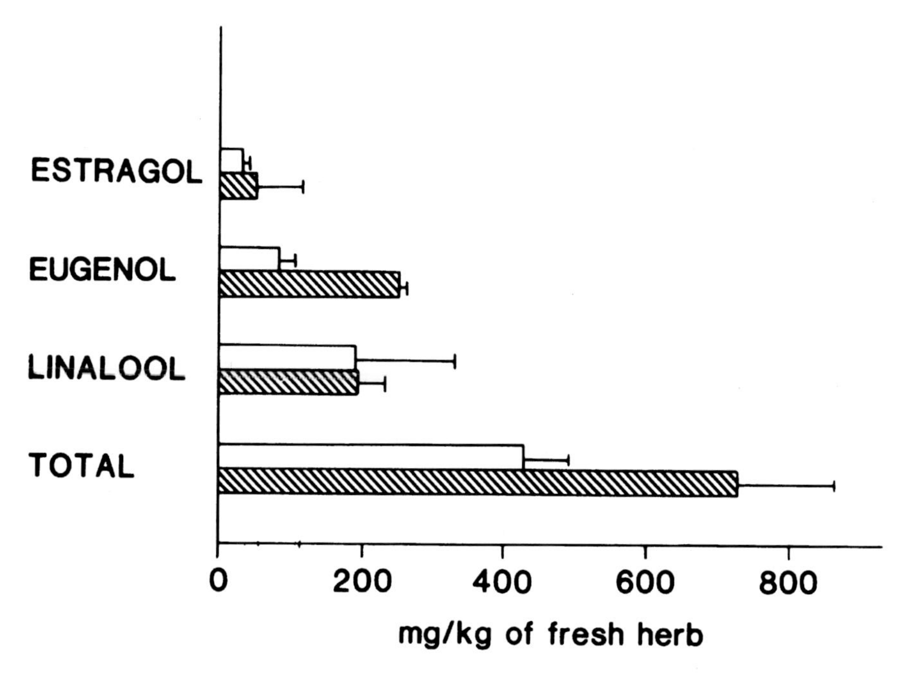

Figure 2.2 Effect of cultivation in open field vs. under cover on aroma compounds from basil produced in Puumala. White column indicates open field and striped column under cover (Nykänen 1986).

deep flow technique. The composition of the nutrient solution suitable for basil is recently studied (Lee *et al.*, 1993, Takano and Yamamoto 1996).

Water

Basil requires a continuous water supply and it is intolerant to water stress at all stages of development. In countries with a warm climate irrigation is an absolute precondition of basil cultivation and irrigation systems are an integrated part of the whole production system. The results of a detailed greenhouse experiment showed that the leaf area and the leaf dry weight were significantly decreased when the plants were subjected to even mild water stress. Mild and moderate water stress (irrigation once every 48 and 72 hours) resulted in a 22% and 41% decrease in leaf dry weight and 15–23% decrease in the leaf area. At the same time, the mild and moderate water stress increased the essential oil content by 87 and 100%, respectively. Although these results suggest the possibility of increasing the oil content by moderate water stress, it is not easily performed in open field cultivation (Simon *et al.*, 1992).

Cultivars

Many basil cultivars are offered by seed companies, but populations arising from seeds are not necessarily genetically uniform. The genus *Ocimum* is characterized by a great variability in both morphology and chemical composition. The ease with which cross-pollination appears leads to a large number of species, subspecies, varieties, and forms. Basil cultivars marketed as seeds are often named according to morphological characteristics or aroma. There are many types of basil preserved locally by farmers/growers, mainly in developing countries. For example in India there are many cultivars which forms the basis for commercial selection (Paton and Putievsky 1996).

For fresh basil production and for garden cultivation many different cultivars are offered. Selected cultivars from seeds include:

Sweet basil:	Sweet, Genovese, Large leaf, Lettuce Leaf, Mammoth
Purple basil:	Dark Opal, Purple Ruffles, Red Rubin, Osmin
Cultivars with different fragrances:	Lemon scented, Cinnamon, Anise, Licorice, Camphor, Spicy bush

During the last years an increasing number of studies have dealt with the morphological and chemotaxonomical features of basil varieties available on the seed market (Simon and Reiss-Bubenheim 1987, Morales *et al.*, 1993b, Grayer *et al.*, 1996, Paton and Putievsky 1996). In Finland 17 different seed batches were examined. All of them were called sweet basil, but a great variation was found in the morphological characteristics as well as in their chemical composition (Galambosi 1995). A similar variability was found when 10 Italian commercially available basil cultivars were examined. The chemical analyses of the varieties showed correlations with the morphological characters. Two cultivars with violet leaves were linalool chemotypes, three of those with large leaves were linalool and methylchavicol chemotypes. Three cultivars with medium sized and two cultivars with small leaves belonged to the linalool and eugenol chemotype (Marotti *et al.*, 1996).

For industrial production mainly sweet basil types are used, and the seed production is performed in controlled circumstances. In essential oil-containing plants the chemicals are sometimes more important characteristics than morphological features, which alone are inadequate to describe the taxon. Therefore the chemotype have to be determined before the plant is used for industrial purposes (Grayer *et al.*, 1996, Paton and Putievsky 1996).

Propagation

Sweet basil is generally propagated by seeds. For selection and preservation of a special plant or cultivar, stem cutting propagation can be used, although this method is not economical for commercial purposes. The ripe, fully mature seeds of basil are generally black or dark brown. The weight of thousand seeds is 1.2–1.8 g. The germination of ripe basil seeds is generally rapid. In laboratory conditions at optimum day/night temperature over 80% germination took four days (Putievsky 1983). In field conditions the emergence occured after 7–14 days in Central Europe and after 4–7 days in India. The crop can be established by direct sowing into the field or by raising the seedlings in a nursery and then transplanting them into the field. Sowing and transplanting takes place in the early summer when there is no danger of night frost.

Direct sowing

For direct sowing generally 2–6 kg/ha of high quality seeds are used. The row distance is 30–50 cm, depending on the weed control system. The sowing time in Central Europe is during the second half of April. During this period the moisture content of the soil is favourable for a good start of the growth. The seedbed should be well tilled and even. The depth of sowing is 0,5–1 cm, and therefore the soil is compacted by rolling prior to and after the sowing. Different seeders are suitable for basil sowing, the best are the vegetable seeders.

Transplantation

In dry areas, like India and Egypt, or for fresh market production in the temperate zones among the smaller growers transplantation of seedlings is the usual propagation method. The seedlings are grown in open-field nurseries or in plastic tunnels in the temperate zones. Seeds are placed in rows (10–15 cm row-spacing), 0.1–0.5 kg seed is required for one hectare. Germination starts 3 days after sowing and is finished in about 10 days. Lateral branching and growth may be encouraged by topping when the plants are about 12 cm high. Regular irrigation is necessary. The seedlings are ready for transplantating after 4–6 weeks. At that stage they are 10–15 cm high with 4–6 pairs of leaves.

The seedlings ready for planting are removed from the nursery and protected from the sun and desiccation. Transplantating is carried out manually or by using vegetable transplanters. Due to the intensive growth the recommended spacing for seedlings in warmer areas is 40–60 cm, with a plant distance of 20–40 cm. In Central Europe the spacing is a little bit smaller, row spacing of 40–50 cm, with 25–30 cm plant distance (Wijesekera 1986, Hornok 1992).

Population density

The optimum population density is partly dependent on the final use of the cultivated basil plants. A high density can be used if compatible farm equipment is available for mechanical cultivation and harvesting. Rows 30–90 cm apart with plants spaced every 25–40 cm are commonly used. When planning the population density and the biomass production the grower has to take into consideration the nutrient availability for the individuals in the plantation. The competition between individuals in a population with a low nutritional supply was found to be intensive and the plants developed more roots than the individuals in a population with a high nutritional level (Morris and Myerscough 1991). According to experiences from Israel the optimum plant density for oil and dry herb production in field conditions is about 15–17 plants/m^2 and for fresh plant production about 8–14 plants/m^2.

Nutrient Supply

An important factor affecting the quantity and quality of the harvested basil yield is to find the optimum level of fertilization. Fertilization experiments carried out with sweet basil focused on the determination of the quantity and the optimum ratio of the main mineral elements. Additionally the nitrogen side dressing was studied in connection with frequent harvests. Special attention was paid to the effects of fertilization on the content and composition of the essential oil. The level and the amount of fertilization naturally depends on the soil, but most papers do not give full details regarding the composition of the soil.

Basil responds well to moderate fertilization. This could be seen in some fertilization experiments, which were carried out in a temperate climate, where the growing season is quite short (April/May–September). This means that the fertilizers are applied during a short time period. In a German study the optimum herb yield of basil was reached by using a compound fertilizer of NPK = 104–12–73 kg/ha. Dividing the total fertilizer application into basic fertilization and top-dressing gave the best results (Weichman 1948).

In a three-year experiment (1973–1975) carried out in Poznan, Poland, increasing N-doses (0–200 kg/ha) increased the total dry herb yield significantly. Without N-fertilization, at P-60 and K-120 kg/ha level the herb yield was 2.27 t/ha. 200 kg/ha nitrogen increased the yield by 44%. Higher phosphorus and potassium doses, 80 and 160 kg/ha respectively, did not increase the basil yield significantly (Czabajski, 1978).

Similar results were achieved in a Hungarian study. On a sandy soil of low humus and nutrient content, the best fresh and dry herb yield (40 t/ha and 7.3 t/ha, respectively) from two harvests were achieved with medium doses of NPK elements (120–100–100 kg/ha). The essential oil yield was highest (40.5 kg/ha) at this fertilization level as well. No beneficial effects of higher doses were found. Increasing

N-doses gave a considerable rise both in the fresh and dry yield of sweet basil (Wahab and Hornok 1981).

In a cooler climate the optimum fertilizer application in respect to the herb yield proved to be rather low compared to the amounts used in warmer countries. This was one of the results of fertilization experiments made in southern Finland, reported by Hälvä and Puukka (1987). The optimum basic fertilization was the application of NPK 40–16–68 kg/ha. In addition basil received a benefit from the N-top dressing 80 kg/ha. Due to the cold weather, the fresh yields were quite low (Figure 2.3). Additionally, in cool and moist conditions the N-top dressing increased the susceptibility to fungus-diseases (*Pythium* sp., *Fusarium* sp., *Sclerotinia sclerotiorum*). The heavier fertilized the crop-stand, the more infected plants there were in both experimental years (Hälvä 1987).

Due to the longer growing season and the accelerated growth in countries with a warmer climate about 3–5 sequent harvests of basil can be obtained each year. In this case the higher phytomass requires a higher nutrient supply and an additional N-dressing after the harvests are important in order to enhance the growth. The different forms of N-dressings may affect the plant growth in different ways.

In a greenhouse pot experiment at Purdue University, USA, the effect of two nitrogen forms (NO_3 and NH_4) were studied on the growth and essential oil content of sweet basil. Ammonium nitrogen decreased the plant height and stem dry weight, but the leaf dry weight was not affected. Ammonium nitrogen also decreased the essential oil content by 28% and relatively increased the amount of sesquiterpene components in the oil (Adler *et al.*, 1989).

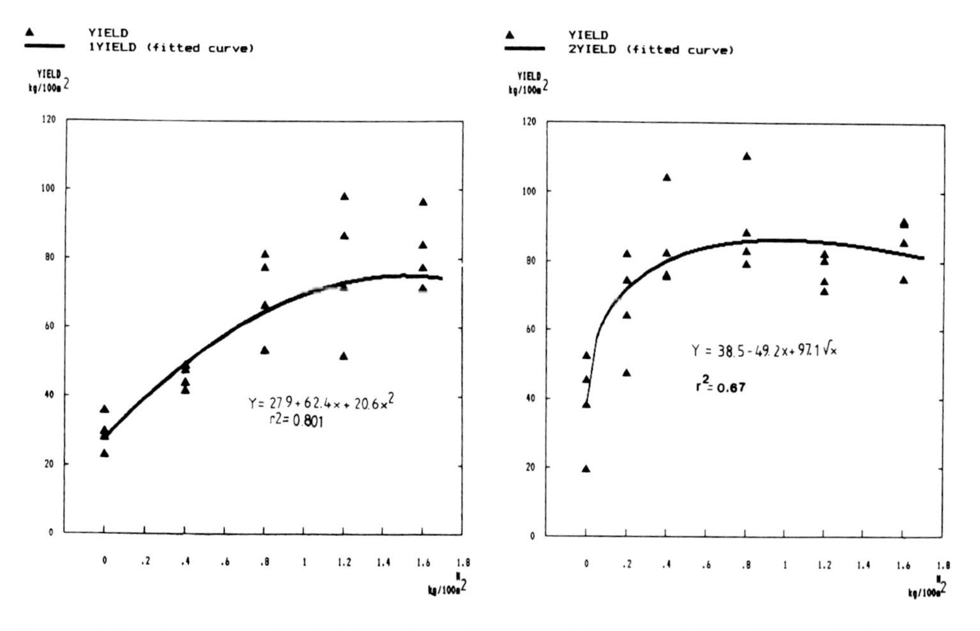

Figure 2.3 Response of basil yield to N-fertilization in 1984 (a) and 1985 (b) (Hälvä and Puukka 1987).

The results of another fertilization study in Italy confirmed the above mentioned Polish, Hungarian and American experiences. According to the results of Tesi *et al.* (1995), "Genovese" sweet basil, grown in greenhouse seemed to be sensitive to high concentrations of fertilizers in the irrigation water. At 500 kg/ha N doses the leaf area decreased 42%, compared to a N dose of 100 kg/ha. The application of different ratios of N, P_2O_5, and K_2O showed that a ratio of 1:1:2 gave the best growth result. Further it was shown that complete fertilization is better than only nitrogen fertilization (Table 2.1). Useful results of these experiments for the fresh herb producers are that slow release fertilizers (ammonium sulphate and sulphonitrate) gave an important reduction in the nitrate content of the fresh plants (more than 50%). This was not seen when using calcium nitrate. Hälvä and Puukka (1987) reported that using calcium nitrate as N-top dressing, 160 kg/ha N-fertilization increased the nitrate content of the fresh herb 15 times.

The practical fertilization advice for basil found in different herb cultivation manuals is based partly on the above mentioned research results, and partly on local soil types, the cropping history and the production systems existing in each country. For example according to Hornok (1992) the Hungarian fertilization advice includes three applications of fertilizers: basic fertilization in the autumn N = 40–60 kg/ha, P = 60–80 kg/ha and K = 120–140 kg/ha are recommended. Start of the fertilization in the spring at the time of soil preparation with N = 40–60 kg/ha and P = 18–20 kg/ha. Foliar N-fertilization is applied after the foliage cuttings, in doses of 60–70 kg/ha.

In the United States the suggested ratio of N–P_2O_5–K_2O elements is 1:1:1, with a N-dose of 230–300 kg/ha, as broadcasts and plowdown. Nitrogen side-dressing at rates of 50–75 kg/ha are recommended after each harvest (Simon 1995).

In India, in the practical cultivation, NPK = 20–40–40 kg/ha basic fertilization is recommended before the soil preparation. 40 kg/ha nitrogen fertilization is applied as top dressing in two equal doses (Srivastava 1980).

In Egypt 35–40 t/ha organic manure and 35 kg/ha P are applied as basic fertilization. As top dressing, 35 kg/ha nitrogen is applied two times, four and seven weeks after transplantating. In addition 35 kg/ha N is applied after each harvest (Shalaby 1996).

Table 2.1 Growth of sweet basil as affected by increasing doses of a 20–20–20 complete fertilizer (Tesi *et al.*, 1995)

Fertilizer Dose g/l	*Nitrogen Supplies kg/ha*	*Plant Height cm*	*Fresh Weight g/plant*	*Leaves Area cm^2/plant*
0	—	0.46 c	6.1 d	20 c
1	100	2.90 a	13.8 b	120 a
2	200	2.70 a	15.4 a	111 a
3.5	350	2.50 a	13.6 b	96 a
5	500	1.90 b	10.3 c	70 b

Mean separation within columns by SNK test at p = 0.05.

High doses of fertilizers are applied combined with regular irrigation in the cultivation of basil in Israel. The ratio of the main mineral element is about 2:1:1. Phosphorus (P_2O_5) and potassium (K_2O) are applied in doses of 100 kg/ha and 50 kg/ha, respectively, and an ammonium nitrate fertilizer is applied immediately before sowing and after each harvest, 250 kg/ha on each occasion (Putievsky and Basker 1977).

Weeding

The presence of weeds among fresh or dry basil leaves decreases the quality of the final product. Therefore weed control is an important part of the basil production systems. The weed problem in high density field populations exists mainly until the first harvest, later the plants cover the field and suppress the growth of the weeds. The weed control is generally carried out mechanically, since relatively few herbicides are suitable for basil cultivation.

From Israel the pre-emergence application of 60% Neburon herbicide was reported in experimental cultivation. It was sprayed prior to germination at 2 kg/ha doses (Putievsky and Basker 1977). Presently in practical cultivation Diphenamid (2 kg/ha) or Neburon (1.2 kg/ha) are used for pre-emergence application and Neburon for post-emergence/postharvest application. In Israel Ozadiazon can be used at a dose of 2.5 kg/ha for transplanted basil plants.

Preplant applications of 2.2 kg/ha Napropamide provided weed control only for two weeks, but was subsequently ineffective against heavy infestations of *Portulaca oleracea* (Ricotta and Masiunas 1991).

Only experimental results are available in Egypt for Terbacil and Bentazone as post-emergence herbicides in sweet basil culture. The preparations were spread 20 days after the transplantating of the seedlings. Terbacil applied at 0.96 kg active ingredient/ha provided satisfactory weed control for both annual and perennial weeds. The effect approximated that of manual weeding. At the same time Bentazone sprayed in a dose of 0.84 kg/ha was less effective. No significant differences in essential oil content or composition of sweet basil were observed when comparing the different weed control treatments (El-Masry *et al.*, 1995).

In many countries basil sold for consumption must be grown without herbicides. Therefore weeding of basil in organic cultivation is becoming more and more important. There are different ways to keep the plantations free from weeds in the herbicide free cultivation. Generally weeding of smaller fields is carried out manually, i.e. by hand hoeing. Weeding can be done by machines in a high density plant population with larger row distances and low distances between the plants.

Organic mulches are suitable for weed control only in smaller areas. Black plastic mulch has become quite a popular choice for weed control, especially in the north European countries. The plastic mulch provides a good weed growth inhibiting effect and the spreading of the plastic mulch can be mechanised. Black plastic mulch increased the fresh and dry yield of basil, due to the incorporation of sun energy which increased the soil temperature under the mulch (Ricotta and Masiunas 1991). In another study plants transplanted into two rows/raised bed produced significantly higher fresh yield in double rows (3854 g/m^2) than in single rows (2501 g/m^2)

(Table 2.2). The use of mulched raised beds seemed suitable for the production of fresh basil, since the harvested yield was cleaner than the yield obtained from cultures without plastic mulch. The production cost for this type of culture is higher than for a normal field production and it reflects on the price of the final product (Davis 1993).

Irrigation

Basil needs a sufficient water supply during almost all stages of development. In a study where the effect of the water supply was studied on the seed yield of basil, it was stated that the average seed yield from irrigated plots was 993 kg/ha compared to 251 kg/ha from dry plots (Cuocolo and Duranti 1982).

The irrigation possibilities, facilities and methods determine the field cultivation systems and the size of the beds irrigated. Canal, flooding and sprinkle irrigation systems are used. In warm countries regular irrigation is carried out at 7–10 days intervals. For example in Israel 350 m^3/ha water is used with 10 days intervals if the field is irrigated by sprinklers. The irrigation intervals are 5 days shorter if the field is irrigated by a drop system.

Irrigation is especially important in the plant nurseries, immediatelly after the transplantating and during the plant emergence after direct sowing. In Egypt flood irrigation is abundant in the Nile valley lands. In India, at the onset of the rain season the rains meet the water requirements of the crop fully until September. After that irrigation may be required once or twice a month (Srivastava 1980). In temperate zones, the distribution of natural rainfall is generally enough during the spring and autumn, but longer dry summer periods without artificial irrigation may cause significant yield losts.

Table 2.2 Effects of in-row spacing and multiple rows per bed on fresh leaf yields of sweet basil, 1990 (Davis, 1993)

Week of	*In-row Spacing (cm)*					*Rows per Bed*		
Harvest	*15*	*23*	*31*	*F-test*	*LSD 5%*	*1*	*2*	*F-test*
	(yield, g/m of Bed)[1]					(yield, g/m of Bed)[1]		
1	232.5	215.4	171.5	*	43.2	162.7	250.3	**
2	357.9	308.2	260.9	*	66.0	264.3	353.7	**
3	297.7	259.3	237.3	ns	231.5	298.0	*	
4	324.5	228.6	309.6	ns	229.0	346.1	**	
5	421.9	383.5	316.0	ns	318.1	429.4	**	
6	317.4	264.2	217.7	*	69.4	239.5	293.4	ns
7	468.3	499.8	458.4	ns	441.3	509.7	ns	
8	482.5	315.3	295.7	*	124.1	302.9	426.1	*
9	346.8	353.9	302.0	ns	311.9	356.6	ns	
Total	3249.4	2828.2	2569.0	**	355.4	2501.2	3263.2	**

[1] Yield in g fresh weight per meter of row length; means of 3 replicates; **, * and ns = significant at $P < 0.01$, 0.05, or nonsignificant, respectively.

Plant Protection

Pests

Compared to other horticultural crops, only a few pests and diseases attack the sweet basil plant. Cultivation manuals from countries in the temperate zone do not offer much data on the pests of sweet basil. Observations of the common bug (*Lygus pratensis* L.) are reported on seedlings after the transplantating (Heeger 1956).

Pests occur more frequently in warmer climates. Heavy infections of *Nematodes* have been reported from Egypt, India and Florida, USA. The infection causes a reduction in the plant growth and oil yield. In addition to crop rotation, different nematicides (aldicarb, carbofuran, bavistin) or oilseed cakes can be used against the most common nematode (*Meloidogyne incognita*) (Haseeb *et al.*, 1988). The phytoparasitic nematode (*Dolichodorus heterocephalus*) significantly reduced the populations of plant parasitic nematodes in common basil culture (Rhoades 1988).

The larvae of some pests may cause serious damage to the plants by sticking to the underside of the leaves, folding them from midrid lengthwise and webbing them. The infected leaves finally fall off. Ocimum leaf folder (*Syngamia abruptalis* Walker) and some lace bugs (*Monanthia globulifera* Walker) have been reported as pests of different basil species in Thailand (Tigvattnanont 1989, 1990). In the càse of serious damage, spraying of insecticides is necessary.

Diseases

Numerous papers report about the presence and damages of different diseases in basil cultivations. The control of diseases is a common part of the cultivation techniques, mainly in countries with a warm climate, where the plantations are regularly irrigated. In the nurseries the seedlings sometimes have to be protected from common soil pathogenes, like *Pythium, Alternaria and Rhysoctonia* sp. According to Srivastava (1980) the following diseases are well-known in basil fields in India: blight of basil, leaf blight and basil wilt.

Blight of basil is caused by *Alternaria* sp. The disease starts with a chlorotic appearance of the leaves which turn purple and finally black. For control it is recommended to spray the crop with 0.2% Dithane M-45, one to three times.

Leaf blight is caused by *Colletotrichum capsici* (Sy.) Butler et Bisby. The symptomes are similar to those of blight. Older leaves appear to be more susceptible to infection.

Basil wilt is caused by *Fusarium oxysporum.* This disease affects the plant at all stages of growth, particularly in the rain season. It appears initially by wilting of the leaves and the shoot tips on one or two branches, but is soon spread to the whole plant, which finally dies. Occurrence of *Fusarium* has been reported in India and Russia (Dzidzariya and Diorbelidze 1974), in France (Mercier and Pionnat 1982), in Italy (Tamietti and Matta 1989) and in the USA (Wick and Haviland 1992, Davis *et al.*, 1993). For prevention of *Fusarium* infection the seedlings should be dipped in a solution of some effective fungicide, such as Tafasol or Agalol, before transplanting. In Russia, methyl bromide was used as an effective fumigant. The evaluation of basil germplasms in order to find strains resistant to *Fusarium* have started and new resistant varieties are in breeding experiments (Reuveni *et al.*, 1996, 1997).

In addition to the above mentioned diseases, the following species were reported to occur on basil or to cause problems in basil cultivation: *Cercospora ocimicola* (Upadhyay *et al.*, 1976), *Corynespora cassiicola* (Devi *et al.*, 1979), *Alsinoe arxii* (Sridhar and Ullasa 1979), *Erisiphe biocellata* (Sharma and Chaudhary 1981), *Rhizoctonia solani* (Sharma 1981), *Pseudomonas cichorii* (Miller and Burgess 1987) and *Pseudomonas syringae* (El-Sadek *et al.*, 1991).

Mechanization

Almost all phases of the production system can be mechanized. The degree of mechanization can be very different in different production areas, depending on the production size, traditional or economic factors. Generally sowing or planting of seedlings, interrow weeding and harvest is carried out mechanically, although there are places where every phase is made manually. The elements of the production system are adapted to locally available machinery. In large scale commercial harvesting motorized cutters or movers with an adjustable cutting height are used.

Harvesting

Among the many different factors, which influence the quality of the dried or distilled basil yield, one of the most important is the harvesting time. The development of the plant organs containing essential oil in different quantities and composition is a continuous process and to find the optimum harvest date is quite difficult. Several studies revealed differences in the oil content and composition due to leaves of different size and age or due to flowering of the plant.

Bettelheim *et al.* (1993) have found that during the ontogenesis the essential oil content of the leaves stayed at the same level (0.3%), while there were great differences in the oil content of the flowers at different development stages. The highest oil content of the flowers (1.0%) was found at an early development stage (Table 2.3).

Great differences were found in the essential oil content and composition of young and mature leaves. Usually, young leaves had a higher content of essential oil per area unit compared to old leaves. As the leaf size increased, the essential oil content decreased. The oil content of young, 1 cm long basil leaves was 0.47–0.66%,

Table 2.3 Essential oil content in different plant parts of *Ocimum basilicum* (Bettelheim *et al.*, 1993)

	Proportion (%) from Total Fresh Weight of		*Essential Oil Content %*	
Flowering Stage	*Flower*	*Leaves*	*Flower*	*Leaves*
Without flowering	0.0	66.6	0.000	0.366
Early flower initiation	6.2	59.6	1.008	0.366
Late flower initiation	13.2	54.9	0.756	0.352
Flowering	20.1	50.1	0.560	0.318
Full flowering	15.7	46.1	0.422	0.328

while that of the mature, 7 cm long leaves was only 0.13%. The content of the main components also changed when the leaves grew older. The oil of the young leaves was rich in linalol (47–51%), while the methyl chavicol content was higher in the older leaves (41–44%) (Werker *et al.*, 1993) (Table 2.4).

In another study, in which the formation of the biologically active substances of basil was followed during the vegetation period, it was found that in the herb oils the monoterpenes reached their maximum during the flowering stage, while the sesquiterpenes had their maximum during the late flowering and seed ripening stages (Lemberkovics *et al.*, 1993).

In India, the effect of transplanting dates and development stage at harvest on the herb and oil yield were studied (Randhawa and Gill 1995). The maximum herb (4.5–8 t/ha) and oil yields (33–54 l/ha) were measured when french basil seedlings were transplanted at the end of July. Later transplanting decreased the oil content of the leaves (Table 2.5). These results can be understood according to the effects of temperature and day length (Putievsky 1983). There were significant differences in the oil yield depending on the development stage during harvest. The herb and oil yield was maximal at the stage of complete flowering. At the stage of 50% flowering the average herb yield was 4.0 t/ha and the oil yield was 26 kg/ha, at 100% flowering stage the corresponding figures were 5.6 t/ha and 37 kg/ha.

There were great differences in the quality and quantity of the basil oil harvested by different methods in Australia (Bonnardeaux 1992). A linalol-methyl chavicol chemotype of *Ocimum basilicum* cv. "Large green" was grown and harvested four times a year. The highest oil yield (91.6 l/ha/year) was obtained when the plants were harvested with long, 12–20 cm flower spikes. Plants harvested with short, 3–11 cm flower spikes, gave an oil yield of only 66.9 l/ha/year. Distillation of the whole plant gave an oil with a high percentage of methyl chavicol (26–32%), whereas the flower spikes produced an oil rich in linalol (43–59%).

When the optimum conditions for mechanical harvest were studied, it was found that the height of the plants at the harvest time influenced the fresh biomass obtained (Table 2.6). If the plants were cut when they were 25 cm high, the fresh yield decreased 25%, but no changes were observed in the dry leaf and oil yield. The

Table 2.4 Essential oil in proximal and distal parts of young and mature sweet basil leaves (Werker *et al.*, 1993)

	Main Component in Essential Oil (%)						
Leaf Part	*Essential Oil Content (%)*	*1,8-Cineole*	*Linalool*	*β-Caryophyllene*	*Methyl Chavicol*	*Methyl Eugenol*	*Eugenol*
Young leaves (1 cm)	—	—	—	—	—	—	—
Proximal	0.66	6.7	51	0.6	20	0.9	3.3
Distal	0.47	3.1	47	2.1	30	0.3	8.2
Mature leaves (7 cm)							
Proximal	0.12	3.2	41	1.0	41	0.3	7.8
Distal	0.14	3.6	39	1.1	44	0.6	5.3

Table 2.5 Effects of transplanting dates and harvesting stages on essential oil production (Randhawa and Gill, 1995)

	Plant Tissue								
	Herb			*Leaf*			*Inflorescence*		
Treatment	*1988*	*1989*	*Mean*	*1988*	*1989*	*Mean*	*1988*	*1989*	*Mean*
Transplant date					(% oil)				
July 1	0.86	0.66	0.76	1.98	1.17	1.57	1.28	1.19	1.23
July 15	0.80	0.66	0.73	1.99	1.19	1.59	1.46	1.34	1.40
July 30	0.73	0.71	0.72	1.62	1.48	1.55	1.45	1.26	1.35
August 15	0.54	0.71	0.62	1.23	1.27	1.25	1.28	1.18	1.23
August 30	0.84	0.61	0.72	1.21	1.17	1.19	1.16	1.13	1.14
Harvesting stage									
Vegetative	0.86	0.75	0.80	1.53	1.26	1.39	1.31	1.23	1.27
50% Flowering	0.77	0.62	0.69	1.64	1.19	1.41	1.36	1.25	1.30
100% Flowering	0.62	0.64	0.63	1.66	1.31	1.48	1.31	1.18	1.24
$LSD_{0.05}$[1]									
Transplant date	0.11	ns		0.10	ns		ns	0.06	
Harvesting stage	0.05	0.03		0.05	ns		ns	ns	
Interaction	ns	0.08		ns	ns		ns	ns	

[1] $LSD_{0.05}$ = least significant difference; P = 0.05; ns = nonsignificant; analysis on square root transforms data.

cutting height of the plants had no effect on the proportions of the main compounds of the oil (Putievsky *et al.*, 1989).

The maximum oil content of several basil species was found at the stage of 50% seed ripening. For example, a methylchavicol type basil, *Ocimum basilicum* var. *glabratum* was grown from seedlings and harvested at different phenological stages. The maximum herb yield was obtained between the 100% flowering and initiation of seed set stage. At this stage, which appeared 150–180 days after the planting, the fresh plant weight was at its maximum, i.e. 930 g/plant. The maximum essential oil content (3.0%) was found later, 210–240 days after the planting, at 50% seed set and seed ripening stage (Gupta 1996).

Table 2.6 The effect of harvest height on yield and essential oil composition of sweet basil (Putievsky *et al.*, 1989)

Harvest Height (cm above the ground)	*Fresh Yield g/m²*	*Flowering Stage %*	*Dry Leaves g/m²*	*Essential Oil in Dry Leaves cc/m²*	*Main Component in the Essential Oil at the 3rd Harvest (%)*	
					Linalool	*Methyl Chavicol*
10	4900	15	455	6.6	47.8	28.2
17	4238	20	452	6.7	48.7	34.6
25	3801	40	436	7.3	50.6	27.2

From the above mentioned examples it is clear, that the phenological and physiological stage of the plants have a great importance in determination of the optimum harvest time either of dry or fresh herb yield or the essential oil. It also have its own impact on the quantity and quality of the basil oil. The harvest time and the number of harvests depend on the ecological conditions of the plantation in question. Theoretically the distilled oil from each harvest has its own qualitative parameters, which have to be determined after the distillation in the laboratory.

In the temperate climate in Central Europe the growing season usually lasts from April to September. In general two harvests are possible. The first harvest is collected when the plants are in full flowering, which takes place in July, and the second harvest is collected in September, before the onset of night frosts.

In countries with a warmer climate, where the vegetation period is longer and the temperature higher, the plants grow faster and makes it possible to obtain several harvests during the same vegetation period. In the northern plain areas in India, sowing takes place in the spring, during April-May, and the seedlings are transplanted after a month. During the growing season four-five harvests are collected. In Egypt the sowing and transplanting time is from March to early April. The plants are harvested at the beginning of the flowering, and successive cuttings are made with 2–2.5 months intervals. Directly sown plantations usually provide four harvest times, and plantations consisting of transplanted plants give three harvests.

In Israel seeds are sown directly into the field during April and the plants are harvested for dry leaf yield prior to the appearance of flowers. For essential oil distillation plants are harvested in full bloom. Basil is generally harvested three or four times between June and December.

Smaller basil fields generally used for the production of fresh basil are harvested manually with the help of hand sickles. The harvested material is immediately transferred to a cooling system, where the plants are stored at a temperature of 12–14°C. After that the plants are packed according to size and weight transported using cooling systems (air, cars).

Post-harvest Processing

Drying

The quality, colour and aroma, of the final product, is affected by the drying and other postharvest processes. The main purpose of the drying process is to reduce the moisture content of the plant material to a safe limit of 8–10%. The time between the harvest and the drying should be as short as possible. The leaves become dark when exposed to open air after the harvest. The leaves should be washed and cleaned before drying in order to remove weeds and other contaminants.

Nykänen and Nykänen (1987) have compared the content of essential oil in fresh basil herb and in the corresponding samples after drying at room temperature and storing in closed dark glass bottles. The results indicated that the total aroma content was reduced by 43% during drying. In order to minimize the loss of volatile compounds in basil leaves and flowering tops, the material should be dried at maximum 40°C.

Two basic drying systems exist:

1. Natural drying in the shade. Due to economic reasons this way of drying is used in the developing countries, although it has the disadvantage of possible contamination by micro-organisms.
2. Artificial drying by warm air. Due to the controlled conditions of this drying process, it is suitable for industrial production. In production countries and areas different mechanical drying systems exists, like plate-chamber driers, conveyor driers etc. (Hornok 1992). Standard quality requirements regarding the volatile oil content and microbiological contamination are easier fulfilled using artificial drying.

In order to fulfill the standard quality requirements, the dried basil raw material is subjected to different post-drying processes, like stalk removing, crumbling, cutting, cleaning , sorting, screening and grinding. All these processes need special machinery.

Storage

The dried basil is stored protected from light and moisture. Generally basil is packed in air-tight or double burlap bags with or without internal polyethylene linings. Depending on the storage conditions some changes may be expected in the oil content, composition or the colour of dried basil leaves.

In a storage experiment sweet basil was dried at 45°C for 12 hours. After three, six and seven months storage the essential oil content and composition was checked by steam distillation and GC-MS. The loss of total essential oil after three, six and seven months storage was 19%, 62% and 66%, respectively. Among the main components the content of methylchavicol and eugenol decreased drastically during drying and storage, while that of linalool and 1,8-cineole increased during the same period (Baritaux *et al.*, 1992).

In another study slight changes were observed during the storage of dried basil leaves. The plants were carefully dried at 35°C and then stored in white paper "sugar bags" in drawers at an alternating room temperature (14–20°C) for three years. The essential oil content was determined after every six months storage. After 27 months storage no changes were observed in the essential oil content (0.5%) and after 3 years storage only slight changes were recorded in the proportions of the oil components (linalool 56.0% and 28.7%) (Svoboda *et al.*, 1996).

Colour, sensory qualities and storage characteristics of freeze-dried and air dried basil were studied by Pääkkönen *et al.* (1990). Basil is seldom freeze-dried, because of the high cost of this method. The freeze-dried basil exhibited an intensive green colour and the air-dried basil a brown colour. After nine months storage in the light or at a raised temperature (35°C) the colour of the freeze-dried basil was altered only slightly. At room temperature the intensity of odour and taste of the freeze-dried basil was better preserved in vacuum and nitrogen-filled packages than in paper bags. The higher temperature was generally unfavourable for the freeze-dried product. The study demonstrated that the quality of air-dried basil could be maintained for 2 years in air-tight packages at room temperature.

Oil Distillation

Depending on the harvested plant part, two grades of oil are obtained from basil, i.e. flower oil and herb oil. The highest content of essential oil appears at the flowering stage, when the plant is cut with 12–20 cm long flower spikes. The oil produced from flowers alone has a superior note and a higher price (Bonnardeaux 1992).

The distillation of basil is generally carried out in stable and intermittently operated distilleries. The distillation process used for basil is the same, which is commercially used for other spices and essential oil plants (Wijesekera 1986, Denny 1995). The distillation time is 1–1.5 h. Usually fresh matter is used for the distillation, but it can also be done with semi-dry or dry matter as well. In India, it is recommended to allow the harvested crop to wilt in the field for 4–5 hours in order to reduce the moisture content. This facilitates the handling of the material and the packing of the distillation apparatus. However, the freshly cut herb should not be exposed to the sun for prolonged periods of time since this has adverse effects on the quality of the oil.

In India, basil fields for commercial production are harvested five times during the growing season. Four times a floral harvest is collected and finally the whole plant. The essential oil content of the floral parts is about 0.4%, of the whole plant it is 0.1–0.25%. The floral oil yield is 12–13 kg/ha and the whole plant oil yield is 18–22 kg/ha, due to the high fresh yield (Srivastava 1980).

In Israel basil is harvested 3–4 times at the flowering stage, when 50% of the plants have flowers. Any delay of the harvest decreases the regrowth of the plants. The average fresh yield is 75 t/ha/year. It is distilled immediately after the harvest and the essential oil yield is 120–140 kg/ha.

It may be emphasized that both the herb yield and the oil content vary greatly depending on the fertility of the soil, harvesting procedures, as well as seasonal conditions. Bright sunny weather, immediately preceeding the harvest, increases the oil content while cloudy or rainy weather decreases it. Also the interval between the last irrigation and harvesting time is very important. If the interval is longer the oil content increases.

Seed Production

The great popularity of basil sets a continuing demand for seeds and thus for seed production. The seed production is carried out by specialized growers and farms since it requires special biological knowledge and technological skills. Seed production is generally performed on the basis of contracts between industrial end-users and seed companies. Commercial cultivars in each country are generally accepted after examination over 3 to 4 seasons and only the superior plants are propagated. Once selected, production of seeds is necessary for propagation of those special cultivars. There are national and international standards for the quality requirements for basil seeds marketed commercially.

The quality evaluation of seeds suitable for propagation is regulated by the rules of the International Seed Testing Association (Anon 1976). These regulations prescribe the laboratory conditions for basil seeds. The germination of basil seeds is carried out in the dark, at a temperature of 20–30°C, during 14 days and the seeds are placed between papers or on the top of the paper. According to the Hungarian Standards for medicinal and aromatic plant seeds (MSZ 6387–87) first and second quality basil seeds must have a purity of 96 and 98%, respectively, and a germination capacity of 70 and 85%, respectively (Anon 1987).

Usually basil seeds are produced in separate fields without cutting the plants for leaf or oil production. In warmer countries, where the vegetation period is longer, one leaf harvest before or after seed formation can be done. The largest quantity of seeds and seeds with the highest germination ratio were collected from plants which were almost dry in the secondary and tertiary stems and in the lower parts of the branches (Putievsky 1993) (Figure 2.4). At this stage the plants were harvested by seed harvesters, the seeds were cleaned and separated according to the quality requirements. Desiccation of the plants by using total weed killers, helps to dry the plants and the seeds can be collected easier by combine. The seed yield of sweet basil ranged between 200–300 kg/ha in Central Europe (Hornok 1978) and 990–1300

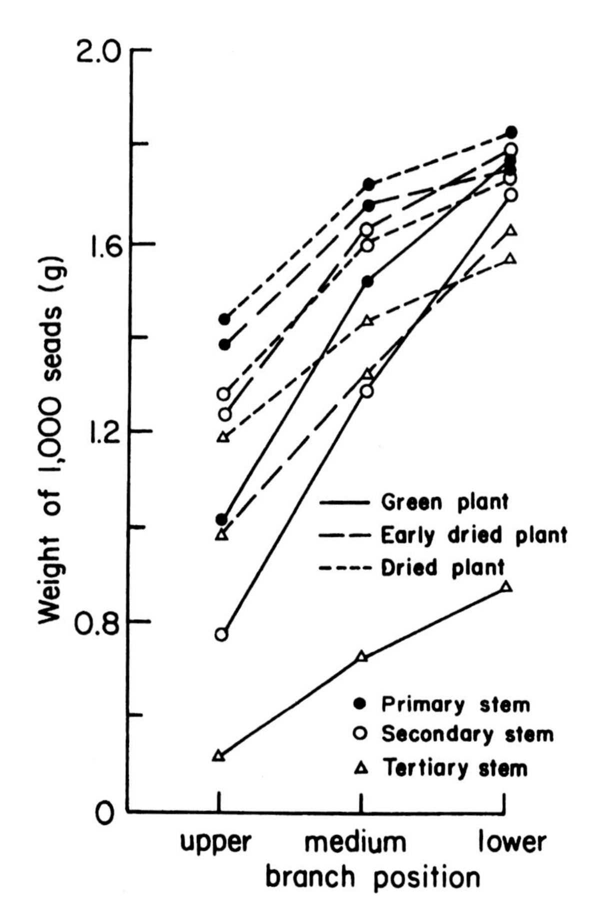

Figure 2.4 Seed yield in sweet basil as affected by branch position, stem-age, and plant maturity (Putievsky 1993).

kg/ha in Italy (Cuocolo and Duranti 1982). In warmer climates the seed yield can be 1500–2000 kg/ha.

Very little is known about the storage of basil seeds. According to practical experience the viability of basil seeds is preserved during storage for quite a few years. On the basis of German experiences Priestley (1986) reported that basil seeds can be expected to maintain a high germination capacity for 4–5 years in favourable conditions of commercial storage. According to own unpublished results, the germination capacity of commercially available seeds of Hungarian origin was 54% and 42%, respectively, after 10 and 12 years storage. Lemon basil seeds produced in a plastic house in Finland had a germination capacity of 81–87% after 6–7 years storage.

UTILIZATION FORMS OF BASILS

Basil is one of the most popular and useful culinary herbs due to its delicate aroma and fragrance. The cultivation methods and production systems vary depending on the part of basil used (fresh or dry leaf, essential oil, seed), on the processing for industrial purposes and on the climatic conditions where it is grown. The main utilization forms of different basil types are fresh basil, frozen, dried, essential oil, medicinal and other uses.

Use of Fresh Basil

The fresh aromatic leaves are used as flavourings or spices in sauces, stews, salads, pickled vegetables, vinegar, aromatic oils as well as in "Bouquet garni". The leaves and flowers of differently coloured cultivars are preferred in restaurants for decoration of the food. In a questionnary in the Los Angeles area, California, basil was the most popular herb among more than 500 restaurants. Of all types of restaurants 79% used basil (Brown 1991).

For fresh use, the most widely used basil types are the large leaf, highly aromatic, so called French or Italian basils, but several other types which differ in leaf size as well as other species with different shape and aroma have more and more commercial importance, for example lemon basil and cinnamon basil. The fresh basil leaves and strigs are sensitive to senescense and therefore the transport between the producer and the final user should be as quick as possible. Otherwise special post-harvest processing, i.e. cooling from harvest up to the consumer, controlled air or special packing, is needed to preserve the quality.

The most advanced production technology for fresh cut basil is the nutrient film techique (NFT), which is also used for the production of fresh salads. Highly automatized and controlled greenhouse production systems can produce standard quality herb all the year round.

Use of Frozen Basil

Frozen basil is used in the same way as fresh basil. Frozen material has the advantage over fresh in that it is not dependent on cultivation periods. The use of frozen basil has become more and more popular in Europe and North America.

The cultivation methods used to produce basil for consumption as fresh or frozen are very similar. In warm countries basils are grown in small areas outdoors or in plastic houses, and are sold on the local market. To ensure a continuous supply, planting dates are normally staggered. Commercial or large scale production of basil generally takes place in greenhouses or plastic houses, where the plants are grown in peat or artificial soils. During the winter time, in the northern and temperate zones, additional light and heat is necessary for the normal photosynthetic activity of the plants.

Use of Dried Basil

Dried foliage is the most important utilization form of basil after basil oil. Preparation of dried basil is practiced in the home gardens, mostly for use in the own kitchen. Large-scale production, using highly industrialised production techniques and quality control, provides basil leaves of hiqh quality for the food industry. Dried basil leaves are the most popular herb in the French, Italian, Mexican and Greek cousines, especially in tomato-based recipes. Basil is used to flavour stews, sauces, sausages, and is also used as a seasoning for confectionery products and chartreuse liqueur.

Use of Basil Oil

The essential oil and oleoresin of basil are extensively used in the food industry, including confectionery, baked goods, meat products and liqueurs. The other main consumption area of basil oil is in perfumery. Basil oil is used in perfumes, soaps, shampoos and dental products. In the perfumery the chemical composition of basil oil has a great importance. The oils from different types or different geographical production areas have a different value on the world market.

The effective production systems for the production of high quality dried leaf yield or essential oil of basil are concentrated to the most favourable climatic regions, where the high investment costs of the industrialized production system can be profitable due to the high yields. The elements of these production systems, described in the above subchapter, are a result of long tradition, intensive research and development efforts made by each country.

The essential oil distillation process is the same as used for other species of aromatic plants. The steam for distillation can be supplied by fuel (in modern apparatus) or from burning plants (in old, caldera type distillation). The plants are harvested at the flowering stage and distilled as soon as possible. The quality of the oil (composition and aroma) depends on the variety, season, development stage and distillation technique.

Medicinal Use of Basil

Traditionally basil has been used as a medicinal plant for various ailments, such as headaches, coughs, diarrhea, constipation, warts, worms and kidney malfunction. It is also thought to be an antispasmodic, stomachicum, carminative, antimalarial, febrifuge and stimulant (Wome 1982, Giron *et al.*, 1991). Ethnobotanical surveys report the traditional utilization of basil as a veterinary medicinal plant as well (Baerts and

Lehmann 1991). Basil oil, especially the camphor containing oil, has antibacterial properties.

Other Uses of Basil

Basil seeds

The main function of basil seeds is for propagation. The seed production is a small, specialized area of basil cultivation, which has great importance in the main production areas and also in the international seed market (Hornok 1992, Putievsky 1993). Lately the fatty oil derived from basil seeds have been of some interest in the cosmetic industry (Riaz *et al.*, 1991, Domokos and Peredi 1993). The main fatty acids of basil seed oil are linolenic (43–64%), linoleic (17–31%), oleic (8–13%) and palmitic acid (6–11%) (Angers *et al.*, 1996).

Ornamental

Different cultivars of basil are traditionally used in the Mediterranean area as pot cultures for decoration, aromatization and sentimental purposes. Presently both in the temperate and the nordic climate the popularity of the containerized herbs, including basil has increased (Dumville 1989, Pessala *et al.*, 1996). Wide germplasm evaluations have been carried out in order to find new interesting forms for decorative purposes (Morales *et al.*, 1993b, Simon and Reiss-Bubenheim 1987).

Edible flowers

The flowers with different fragrance and colour (white, pink, violet) are used together with summer vegetables, cheese, fish, butter, oils, etc. The flowers are also used for religious purposes.

MARKET POTENTIAL OF BASIL PRODUCTS

Although basil has a wide popularity, no up-to date figures can be found on the national and international trade statistics. Especially there is lack of information of cultivation areas. The market potential for basil products is summarized below.

Basil Oil

The total world production of basil oils can be estimated to 93–95 t/year of which 55 tons for *Ocimum gratissimum* (800.000 dollars in value), 43 tons for *Ocimum basilicum*, (2.8 million dollars). About 100 kg of oils are produced from *Ocimum canum* (5000 dollars). During the last few years, the most conservative estimate is that the production of basil oil will increase by a few percents yearly (Lawrence 1993).

Basil oils are currently produced in the following countries (the quantities in brackets are tons): India (15), Bulgaria (7), Egypt (5), Pakistan (4.5), the Comoros (4.5), Israel (2), the former Yugoslavia, USA and Madagascar (each 1), Reunion and

Albania (each 0.5), Hungary (0.3) and Argentina (0.2 (Lawrence 1993). In the former Soviet Union the reported production of *Ocimum gratissimum* oil during 1972, 1973 and 1974 was 70, 95 and 141 tons, respectively (Anon 1974).

USA is probably the largest market for basil oil, followed by the European countries, Germany, France, UK and the Netherlands (Robbins and Greenhalg 1979). From Spain a consumption of 5 t/year has been reported (Porredon 1987).

Basil Herb

In contrast to basil oil, there are no available statistics concerning the world production of dried basil herb. A considerable proportion of the world production, particularly in the Mediterranean area, in India and in California, does not enter the international trade, but is rather consumed locally. For example, the domestic dried basil consumption during the 1970s in Hungary was 70–90 t/year and 100–120 t was exported (Hornok 1978).

Available import statistics reveal that USA is one of the biggest consumers of dried basil, due to the increasing popularity of Italian and Mexican style cooking. According to Greenhalg (1979) and Simon (1988) the basil import in 1964 was 19 t, in 1976 412 t and during 1986–1988 1400–1800 t/year. The value of the basil import during 1988 was 2.456.000 dollars. During the period 1972–77 the main suppliers of USA were Bulgaria (71 t in 1972), Hungary (87 t in 1972), Mexico (86 t in 1973) and Egypt (150 t in 1976).

Other main areas for basil import is the European countries. According to the latest market survey, the total basil herb import to Europe was about 830–880 t/year (Anon 1991). France is the biggest importer, 300–350 t/year, followed by the United Kingdom (250 t/year), Germany (200 t/year) and the Netherlands (80 t/year). The import of basil to Belgium and Switzerland was 10 and 5 t/year, respectively (Svoboda 1984). The import to Finland during 1982 was 6.6 t (Hälvä, 1985). One of the main suppliers of the Western European countries is Egypt. The production area in Egypt was 1238 ha in 1994 and the following quantities of dried basil were exported: to United Kingdom 200 t, to Germany 150 t, to France 140 t and to the Netherlands 50 t (Shalaby 1996).

Fresh Basil

There is very little information about the fresh basil market, mainly because the fresh plant material is used quickly and locally. Italy is the largest fresh basil producer with 5000 t/year, followed by France (2000 t/year), Israel (500 t/year) and North-Africa (mainly Morocco, about 100–150 t/year) (Bianco 1992). Additionally there is significant production in the USA and Central America. In the north European countries fresh basil is produced hydroponicly. In Finland fresh basil herb was third on the popular herbs list (1995), including 6 millions fresh cut basil herbs (Galambosi 1996).

Production Areas

Due to its great popularity basil is grown all over the world, in the warm and temperate zones. Basil cultivation is practiced in the following countries:

Warm climate:	India, Pakistan, the Comores, Madagascar, Haiti, Guatemala, Reunion, Thailand, Indonesia, Russia (Georgia, East-Caucasus) and South Africa
Mediterranean area:	Egypt, Morocco, France, Israel, Bulgaria, USA (Arizona, California, New Mexico), Italy, Greece and Turkey
Temperate zone:	Hungary, Poland, Germany, Balkan countries, Slovakia

From the available trade statistics and the average basil yields in the different production areas, we tried to estimate for the first time the total cultivation area of sweet basil in the world. As a result of the calculations, the sweet basil areas for oil production is estimated to be about 1700 ha, and 2200 ha for dried herb production. The estimated domestic consumption in the traditional basil consumer countries and other areas, such as Africa requires about 1000–1500 ha field areas. The total growing area of sweet basil is estimated to be about 5000 ha in the world.

REFERENCES

Abu-Zeid, E.N. (1988) *Medicinal plants and their products.* El-Dar El-Arabia Press, Cairo, Egypt.

Adler, P.R., Simon, J.E. and Wilcox, G.E. (1989) Nitrogen form alters sweet basil growth and essential oil content and composition. *HortScience*, **24**, 789–790.

Adnan, S.A.W. and Hornok, L. (1982) Effect of NPK fertilisation on yield and essential oil content of sweet basil (*Ocimum basilicum* L.). *Public. Univ. Hort. Budapest*, **46**, 67–73.

Aharoni, N., Dvir, O., Chalupowicz, D. and Aharon, Z. (1993) Coping with postharvest physiology of fresh culinary herbs. *Acta Hort.*, **344**, 69–78.

Angers, P., Morales, M.R. and Simon, J.E. (1996) Fatty acid variation in seed oil among *Ocimum* species. *JAOCS*, **73**, 393–395.

Anon (1974) Data of the essential oil production of USSR. *Express information of essential oils*, **8**.

Anon (1976) International Rules for Seed testing. *Seed Sci. & Technology*, **4**, 3–49.

Anon (1987) *Gyogy-, Illoolaj-, fuszer- es mezelönövenyek vetömagjai* (Hungarian standards for seeds of medicinal, volatile oil, spice and melliferous plants). MSZ 6387–87.

Anon (1991) *Dry culinary herbs. An overview of selected western European markets.* ITC, Geneva, Switzerland, p. 60.

Baerts, M. and Lehmann, J. (1991) Veterinary medicinal plants of the region of Cretes Zaire-Nil in Burundi. *Annalen Economische Wetenschappen-koninklijk Museum voor Midden-Afrika*, **21**, 133.

Baritaux, O., Richard, H., Touche, J. and Derbesy, M. (1992) Effects of drying and storage of herbs and spices on the essential oil. Part I. Basil, *Ocimum basilicum* L. *Flav. Fragr. J.*, **7**, 267–271.

Basker, D. and Putievsky, E. (1978) Seasonal variation in the yields of herb and essential oil in some Labiatae species. *J. of Hort. Science*, **53**, 179–183.

Bettelheim, Y., Dudai, N., Putievsky, E., Ravid, U., Saadi, D., Katzir, I., Michaelovich, Y. and Cuabi, E. (1993) The influence of flowering and environmental factors on yield components and essential oil in exotic sweet basil (in Hebrew). *Hassadeh*, **73**, 961–965.

Bianco, V.V. (1992) Usual and special vegetable crops in mediterranean countries. *Acta Hort.*, **318**, 65–76.

Bonnardeaux, J. (1992) The effect of different harvesting methods on the yield and quality of basil oil in the Ord River irrigation area. *J. Ess. Oil Res.*, **4**, 65–69.

Brown, S.H. (1991) Culinary herb use in southern California restaurants. *California Agriculture*, **45**, 4–6.

Czabajski, T. (1978) Wplyw wyskich dawek azotu na plon ziela bazylii I czabru. *Wiadomosci Zielarskie*, **7**, 11.

Cantwell, M.I. and Reid, M.S. (1993) Postharvest physiology and handling of fresh culinary herbs. *J.of Herbs, Spices & Medicinal Plants*, **1**, 93–127.

Cuocolo, L. and Duranti, A. (1982) The effects of irrigation and nitrogen fertilizing on the seed yield of basil, cv. Fino Genovese. *Rivista di Agronomia*, **16**, 17–122.

Csedö, K. (1980) *Plantele Medicinale si Condimentare Din Judetul.* Harghita Tipografia, Tirgu Mure, Romania.

Dachler, M. and Pelzmann, H. (1989) *Heil- und Gewurzpflanzen. Anbau – Ernte – Aufbereitung.* Österreicher Agrarverlag, Wien, Austria.

Darrah, H.H. (1972) The basils in folklore and biological science. *The Herbarist*, **38**, 3–10.

David, W. and Steward, K.A. (1986) The potential of NFT for the production of six herb species. *Soils Culture*, **2**, 61–70.

Davis, J.M. (1993) In-row plant spacing and yields of fresh-market basil. *J. of Herbs, Spices & Medicinal Plants*, **2**, 35–43.

Davis, R.M., Marshall, K.D. and Valencia, J. (1993) First report of *Fusarium* wilt in California. *Plant Disease*, **77**, 537.

Denny, E.F.K. (1995) *Field Distillation for Herbaceous Oils.* Denny, Mckenzie Associates, Lilydale, Tasmania, Lilydale.

Devi, L.R., Menon, M.R. and Nair, M.C. (1979) *Corynespora* leaf spot of sweet basil. *Indian Phytopathology*, **32**, 150–151.

Domokos, J. and Peredi, J. (1993) Studies on the seed oils of basil (*Ocimum basilicum* L.) and summer savory (*Satureja hortensis* L.). *Acta Hort.*, **344**, 312–314.

Duke, J.A. and Hurst, S.J. (1975) Ecological amplitudes of herbs, spices and medicinal plants. *Lloydia*, **38**, 404–410.

Dumwille, C. (1989) Containerized herb plants: product development and marketing through garden centre outlets. *Professional Horticulture*, **3**, 31–34.

Dzidzariya, O.M. and Giorbelidze, A.A. (1974) Trials of fumigants against the pathogens of diseases of East Indian basil (In Russian). Sb. Statei po Efirnomaslich. Kulturam. I Efirn. Maslam. Suhumi, 93–97.

El-Masry, M.H., Charles, D.J. and Simon, J.E. (1995) Bentazon and Terbacil as postemergent herbicides for sweet basil and sweet marjoram. *J. of Herbs, Spices & Medicinal Plants*, **3**, 19–26.

El-Sadek, S.A.M., Abdel-Latif, M.R., Abdel-Gawad, T.I. and El-Sakawy, F.S. (1991) Occurrence of leaf blight of basil caused by *Pseudomonas syringae* in Egypt. *Assiut J. of Agric. Sci.*, **22**, 91–109.

Galambosi, B. (1995) Basilika (*Ocimum basilicum* L.). In *Organic Cultivation of Herbs and Medicinal Plants*. Painatuskeskus Oy, Helsinki, Finland, pp. 144–146.

Galambosi, B. (1996) Selviä laajenemispiirteitä yrttien viljelyssä. *Puutarha*, **99**, 350–351.

Giron, L.M., Freire, V., Alonzo, A. and Vaceres, A. (1991) Ethnobotanical survey of the medicinal flora used by the cribs of Guatemala. *J. Ethnopharmacol.*, **34**, 173–187.

Grayer, R.J., Kite, G.C., Goldstone, F.J., Bryan, S.E., Paton, A. and Putievsky, E. (1996) Infraspecific taxonomy and essential oil chemotypes in sweet basil, *Ocimum basilicum. Phytochemistry*, **43**, 1033–1039.

Greenhalgh, P. (1979) *The market for culinary herbs.* Tropical Product Institute. London. G 121.

Gupta, S.C. (1996) Variation in herbage yield, oil yield and major component of various *Ocimum* species/varieties (chemotypes) harvested at different stages of maturity. *J. Ess. Oil Res.*, **8**, 275–279.

Haseeb, A., Pandey, R. and Hussain, A. (1988) A comparison of nematicides and oilseed cakes for control of Meloidogyne incognita on *Ocimum basilicum. Nematropica*, **18**, 65–69.

Hay, R.K.M., Svoboda, K.P. and Barr, D. (1988) Physiological problems in the development of essential oil crops for Scotland. *Crop Res.* (*Hort.Res.*), **28**, 35–45.

Heeger, E.F. (1956) *Handbuch des Arznei- und Gewurzpflanzenbaues.* Deutscher Bauerverlag, Berlin, Germany.

Hornok, L. (1978) *Gyogynövenyek termesztese es feldolgozasa.* Mezögazdasagi Kiado, Budapest, Hungary.

Hornok, L. (1992) *Cultivation and Processing of Medicinal Plants.* Akademia Kiado, Budapest, Hungary.

Hälvä, S. (1985) Consumption and production of herbs in Finland. *J. of Agric. Sci. in Finland*, **57**, 231–237.

Hälvä, S. and Puukka, L. (1987) Studies on fertilization of dill (*Anethum graveolens* L.) and basil (*Ocimum basilicum* L.) I. Herb yield of dill and basil affected by fertilization. *J. of Agric. Sci. in Finland*, **56**, 11–17.

Hälvä, S. (1987a) Studies on fertilization of dill (*Anethum graveolens* L.) and basil (*Ocimum basilicum* L.) III. Oil yield of basil affected by fertilization. *J .of Agric. Sci. in Finland*, **59**, 25–29.

Hälvä, S. (1987b) Studies on production techniques of some herb plants. I. Effect of Agryl P17 mulching on herb yield and volatile oils of basil (*Ocimum basilicum* L.) and marjoram (*Origanum majorana* L.). *J. of Agric. Sci. in Finland*, **59**, 31–36.

Lawrence, B.M. (1992) Chemical components of Labiatae oils and their exploatation. In R.M. Harley and T. Reinolds, (eds.), *Advances in Labiatae Science*, Royal Botanical Gardens, Kew, UK, pp. 399–436.

Lawrence, B.M. (1993) A planning scheme to evaluate new aromatic plants for the flavor and fragrance industries. In J. Janick and J.E. Simon, (eds.), *New Crops*, John Wiley and Sons, New York, USA, pp. 620–627.

Lee, B.S., Seo, B.S., Chung, J. and So, C.H. (1993) Growth and oil content in sweet basil (*Ocimum basilicum* L.) as affected by different hydroponic systems. *J.of Korean Soc. For Hort. Sci.*, **34**, 402–411.

Lemberkovics, E., Nquyen, H., Taar, K., Mathe, Jun, I., Petri, G. and Vitanyi, G. (1993) Formation of biologically active substances of *Ocimum basilicum* L. during the vegetation period. *Acta Hort.*, **344**, 334–346.

Marotti, M., Piccaglia, R. and Giovanelli, E. (1996) Differences in essential oil composition of basil (*Ocimum basilicum* L.) Italian cultivars related to morphological characteristics. *J. Agric. Food Chem.*, **44**, 3926–2929.

Menghini, A., Cenci, C.A., Cagiotti, M.R. and Pagiotti, R. (1984) Bioritmi e produttivita di *Ocimum basilicum* L. in differenti condizioni ambientali. *Ann. Fac. Agr. Univ. Perugia*, **XXXVIII**, 287–296.

Mercier, S. and Pionnat, J.C. (1982) The occurrence in France of basil vascular wilt. *Comptes Rendus des Seaces de l'Academie d'Agriculture de France*, **68**, 416–419.

Miller, J.W. and Burgess, S.M. (1987) Leafspot and blight of basil caused by *Pseudomonas cichorii. Plant Pathology Circular*, Florida Dept. of Agriculture, No. 293.

Morales, M.R., Simon, J.E. and Charles, D.J. (1993a) Comparison of essential oil content and comparison between field and greenhouse grown genotypes of methyl cinnamate basil (*Ocimum basilicum* L.). *J. of Herbs, Spices & Medicinal Plants*, **1**, 25–30.

Morales, M.R., Charles, D.J. and Simon, J.E. (1993b) New aromatic lemon basil germplasm. In J. Janick and J.E. Simon(eds), *New Crops*, John Wiley and Sons, New York, USA, pp. 632–635.

Morris, E.C. and Myerscough, P.J. (1991) Self-thinning and competition intensity over a gradient of nutrient availability. *J. Ecol. Oxford: Blackwell Scientific*, **79**, 903–923.

Nykänen, I. (1986) High resolution gas chromatographic-mass-spectrometric determination of the flavour composition of basil (*Ocimum basilicum* L.) cultivated in Finland. *Z. Lebensm Unters Forsch* **182**, 205–211.

Nykänen, I. (1989) The effect of cultivation conditions on the composition of basil oil. *Flav. Fragr. J.*, **4**, 125–128.

Nykänen, L. and Nykänen, I. (1987) The effect of drying on the composition of the essential oil of some Labiatae herbs cultivated in Finland. In M. Martens, G.A. Dalen and H. Jr. Russwurm, (eds.), *Flavour Science and Technology*, John Wiley & Sons, Chichester, pp. 83–88.

Paton, A. and Putievsky, E. (1996) Taxonomic problems and cytotaxonomic relationships between and within varieties of *Ocimum basilicum* and related species (Labiatae). *Kew Bulletin*, **51**, 509–524.

Pessala, R., Hupila, I. and Galambosi, B. (1996) Yield of different basil varieties in pot culture indoor. *Drogenreport*, **9**, 16–18.

Pogany, D., Bell, C.L. and Kirch, E. (1968) Composition of oil of sweet basil (*Ocimum basilicum* L.) obtained from plants grown at different temperatures. *P. & E.O.R.*, 858–865.

Porredon, T.A. (1987) La coltivazione delle piante medicinali e aromatiche in Spagna. *Economia Trentina*, **3**, 78–90.

Priestley, D.A. (1986) *Seed Aging. Implications for Seed Storage and persistence in the Soil.* Comstock Publ. Ass. Ithaca and London, UK.

Putievsky, E. and Basker, D. (1977) Experimental cultivation of marjoram, oregano and basil. *J.Hort. Sc.*, **52**, 181–188.

Putievsky, E. (1983) Temperature and day-length influences on the growth and germination of sweet basil and oregano. *J. Hort. Sci.*, **58**, 583–587.

Putievsky, E., Ravid , U., Dudai, N., Zuabi, E., Michaelovich, Y. and Saadi, D. (1989) The influence of harvesting height on yield components of aromatic plants. *Hassadeh*, **69**, 1421–1422, 1424, 1429.

Putievsky, E. (1993) Seed quality and quantity in sweet basil as affected by position and maturity. *J.of Herbs, Spices & Medicinal Plants*, **2**, 15–20.

Pääkkönen, K., Malmsten, T. and Hyvönen, L. (1990) Drying, packaging, and storage effects on quality of basil, marjoram and wild marjoram. *J. of Food Science*, **55**, 1373–1377, 1382.

Randhawa, G.S. and Gill, B.S. (1995) Transplanting dates, harvesting stage, and yields of French basil (*Ocimum basilicum* L.). *J. of Herbs, Spices & Medicinal Plants*, **3**, 45–56.

Reuveni, R., Dudai, N., Putievsky, E., Elmer, W.H. and Wick, R.L. (1996) Evaluation and identification of basil germplasm for resistance to *Fusarium oxysporum* f. sp. *basilicum.* Contribution from the Agric. Res. Organization, The Volcani Center, bet Dagan, Israel. No. 1887-E. 1996 series.

Reuveni, R., Dudai, N. and Putievsky, E. (1997) "NUFAR" – a sweet basil cultivar resistant to *Fusarium* wilt caused by *Fusarium oxysporum* f. sp. *basilicum. Hort Sci.* (accepted for publication).

Rhoades, H.L. (1988) Effects of several phytoparasitic nematodes on the growth of basil, *Ocimum basilicum. Annals of Applied Nematology*, **2**, 22–24.

Riaz, M., Khalid, M.R. and Chaudhary, F.M. (1991) Lipid fractions and fatty acid composition of different varieties of basil seed oil. *Pakistan Journal of Scientific and Industrial research*, **34**, 346–347.

Ricotta, J.A. and Masiunas, J.B. (1991) The effects of black plastic mulch and weed control strategies on herb yield. *Hort Science*, **26**, 539–541.

Robbins, S.R.J. and Greenhalgh, P. (1979) *The markets for selected herbaceous essential oils.* Tropical Prod. Institute. London, G 120.

Shalaby, A.S. (1996) Basil production in Egypt. Personal communication.

Sharma, A.D. (1981) Collar rot and blight of *Ocimum* species in India. *Indian J. of Mycology and Plant Pathology*, **11**, 149–150.

Sharma, Y.R. and Chaudhary, K.C.B. (1981) Powdery mildew of *Ocimum sanctum* – a new record. *Indian Phytopathology*, **33**, 627–629.

Simon, J.E. (1993) Basil: A source of essential oils. In J. Janick and J.E. Simon (eds.), *New Crops*, John Wiley and Sons, New York, USA, pp. 484–489.

Simon, J.E. (1995) Basil. New Crop Factsheet. Internet http://www.hort.purdue.edu/new crop.

Simon, J.E. and Reiss-Bubenheim, D. (1987) Characteristics of American basil varieties. In J.E. Simon and L. Grant (eds.), *Proceedings of the Second National Herb Growing and Marketing Conference.* July 1987, Indianapolis, Indiana, USA, pp. 48–51.

Simon, J.E. (1988) Aromatic and medicinal plants in the United States. *Compte – Rendus des 2e Recontres Techniques et Economiques Plantes Aromatiques et Medicinales*, December 1988, Nyons, France, pp. 66–86.

Simon, J.E., Reiss-Bubenheim, D., Joly, R.J. and Charles, D.J. (1992) Water stress-induced alterations in essential oil content and composition of sweet basil. *J. Ess. Oil Res.*, **4**, 71–75.

Skrubis, B. and Markakis, P. (1976) The effect of photoperiodism on the growth and the essential oil of *Ocimum basilicum* (sweet basil). *Econ. Bot.*, **30**, 389–393.

Sörensen, L. and Henriksen, K. (1992) Effects of seed rate, plastic covering, and harvest time on yield and quality of Danish grown basil (*Ocimum basilicum*). *Danish Journal of Plant and Soil Science*, **96**, 499–506.

Sridhar, T.S. and Ullasa, B.A. (1979) Scab of *Ocimum basilicum* – a new disease caused by *Elsinoe arxii* sp. *nov.* from Bangalore. *Current Science*, **48**, 868–869.

Srivastava, A.K. (1980) French basil and its cultivation in India. *Farm Bulletin No. 16*. Central Institute Medicinal and Aromatic Plants, Lucknow, p. 15.

Svoboda, K.P. (1984) Culinary and medicinal herbs. The West Scotland Agricultural College, Ayr. Technical note No. 237, p. 12.

Svoboda, K.P., Galambosi, B. and Hampson, J. (1996) Influence of storage on quality and quantity of essential oil yield from twenty herb species. In *Book of Abstracts*, 0 6–2, 27th International Symposium on Essential Oils, September 1996, Vienna, Austria.

Takano, T. and Yamamoto, A. (1996) Effect of anion variations in a nutrient solution on basil growth, essential oil content, and composition. *Acta Hort.*, **426**, 389–396.

Tamietti, G. and Matta, A. (1989) The wilt of basil caused by *Fusarium oxysporum* f. sp. *basilicum* in Liguria. *Difesa delle Plante*, **12**, 213–220.

Tesi, R., Chisci, G., Nencini, A. and Tallarico, R. (1995) Growth response to fertilization on sweet basil (*Ocimum basilicum* L.). *Acta Hort.*, **390**, 93–96.

Tigvattnanont, S. (1989) Studies on the bionomics and local distribution of some lace bugs in Thailand. I. *Monanthia blobulifera* Walk. (Hemiptera: Tingidae). *Khon. Kaen. Agric. J.*, **17**, 333–334.

Tigvattnanont, S. (1990) Studies on biology of the *Ocimum* leaf folder, *Syngamia abruptalis* Walker and its host plants. *Khon. Kaen. Agric. J.*, **18**, 316–324.

Upadhyay, D.N., Bordoloi, D.N., Bhagat, S.D. and Ganguly, D. (1976) Studies on blight disease of *Ocimum basilicum* L. caused by *Cercospora ocimicola* Petrak et Ciferri. *Herba Hungarica*, **15**, 81–86.

Wahab, A.S.A. and Hornok, L. (1982) Effect of NPK fertilization on *Ocimum basilicum* yield and essential oil content. *Kertesz̧eti Egyetem Köz̧lemenyei*, **45**, 65–73.

Wees, D. and Stevart, K.A. (1986) The potential of NFT for the production of six herb species. *Soilless Culture*, **2**, 61–70.

Weichan, C. (1948) Der Gehalt an ätherischen Öl bei Aromatischen Pflanzen in Abhängigkeit von der Dungung. *Pharmazie*, **3**, 464–467.

Werker, E., Putievsky, E., Ravid, U., Dudai, N. and Katzir, I. (1993) Glandular hairs and essential oil in developing leaves of *Ocimum busilicum* L. (Lamiaceae). *Ann.Bot.*, **71**, 43–50.

Wick, R.L. and Haviland, P. (1992) Occurrence of *Fusarium* wilt of basil in the United States. *Plant Disease*, **76**, 323.

Wijesekera, R.O.B. (1986) *Practical manual on: The essential oils industry.* UNIDO, Vienna, Austria, p. 173.

Wome, B. (1982) Febrifuge and antimalarial plants from Kisangani, upper Zaire. *Bulletin de la Societe Royale de botanique de Belgique*, **115**, 243–250.

3. CHEMICAL COMPOSITION OF *OCIMUM* SPECIES

RAIMO HILTUNEN

Department of Pharmacy, P.O Box 56, FIN-00014
University of Helsinki, Finland

INTRODUCTION

Basil herb (*O. basilicum* L.) contains 0.5–1.5% essential oil of varying composition, about 5% tannins and β-sitosterol. Basil seeds contain planteose, mucilage, polysaccharides and fixed oil that consists of linoleic acid (50%), linolenic acid (22%), oleic acid (15%) as well as 8% unsaturated fatty acids. Basil leaves also contain 0.17% oleanolic acid and a small amount of ursolic acid (List and Hörhammer 1977). Also the leaves and flowers of *O. canum* Sims. contain oleanolic and ursolic acids (Xaasan *et al.*, 1980).

O. sanctum L. herb contains 0.6% essential oil and the seeds contain fixed oil that consists of 66% linoleic acid. Possibly the seeds may also contain alkaloids and mucilage (List and Hörhammer 1977). The essential oil content of *O. canum* Sims. (*O. americanum* L.) herb varies from 0.38% to 0.65%. *O. kilimandscharicum* Guerke is rich in essential oil (2.5–7.6%) with 50–70% camphor as the main compound. The seeds contain fixed oil in which linoleic acid is the main fatty acid (50–70%). Fresh leaves of *O. gratissimum* L. contain quite high amounts of essential oil (3.2–4.1%). Generally leaves of *Ocimum* sp. contain essential oil from 0.5 to 1.4%. In addition to essential oil the herb also contains 2,5-dimethoxybenzoic acid (ocimol) and gratissimin (α-truxillic acid dimethylester). The roots contain ocimol, glucose, galactose, arabinose, β-sitosterol and ocimic acid. The essential oil content of *O. viride* Willd. varies from 2.2% to 2.4% (List and Hörhammer 1977). The dried leaves and flower tops of sweet basil (*O. basilicum*) contain essential oil (ca. 0.08%), and protein (14%), carbohydrates (61%) and relatively high concentrations of vitamins A and C, rosmarinic acid and a flavone named xanthomicrol (Leung and Foster 1996). The essential oil of *Ocimum* species is discussed elsewhere in this book, see the chapter: Essential oil of *Ocimum*.

FIXED OILS AND FATTY ACIDS

The fixed oil from *O. basilicum* and *O. album* seeds were collected from the local market in Pakistan. The seed oil of *O. basilicum* was characterized by a high content of linolenic acid (48.50%) and linoleic acid (21.81%). In the seed oil of *O. album* oleic acid (44.16%) was detected as the main compound (Table 3.1). The yield of fixed oil was 21.4% for *O. basilicum* and 15.5% for *O. album* (Malik *et al.*, 1989).

Fixed oil was obtained by cold pressing from the seed material arising from commercial basil (*O. basilicum*) and the oil yield was 22.5% (d.w.). The oil was very rich in α-linolenic acid (50–63%) and also contained a high amount of linoleic

Table 3.1 Percentage composition of fatty acids in the seed oils of *Ocimum album*, *Ocimum basilicum* and *Ocimum sanctum* and oil yields (Nadkarni and Patwardhan 1952, Malik *et al.*, 1987, 1989)

Fatty Acid	*O. album* [1]	*O. basilicum* [1]	*O. sanctum* [2]	*O. sanctum* [3]
Capric acid	1.30	0.00	0.00	—
Lauric acid	0.78	0.85	2.84	—
Myristic acid	0.68	0.36	1.90	—
Palmitic acid	11.68	9.70	5.54	6.9
Stearic acid	2.33	5.45	3.12	—
Oleic acid	44.16	13.33	6.00	9.0
Linoleic acid	36.36	21.18	59.10	66.1
α-Linolenic acid	0.00	48.50	21.27	15.7
Arachidic acid	2.73	0.00	0.00	—
Oil yield (%)	15.5	21.4	18.19	17.82

[1] Malik *et al.*, 1989
[2] Malik *et al.*, 1987
[3] Nadkarni and Patwardhan 1952.

acid (17–25%). The other fatty acids analyzed were oleic (9–15%), palmitic (6–9%) and stearic acid (2–3%). In this study the fixed oil contained 0.5% of free fatty acids (Domokos *et al.*, 1993).

The fixed oil from *O. sanctum* seeds is characterized by a high content of linoleic acid (59.1–61.0%) and linolenic acid (15.7–21.7%) (Nadkarni and Patwardhan 1952, Malik *et al.*, 1987). *O. sanctum* seeds contained 17.50% fixed oil (v/w dried seeds). The oil contained linoleic acid as the main compound (52.2%). Other reported fatty acids were linolenic acid (16.6%), oleic acid (13.8%), palmitic acid (11.7%) and stearic acid (3.2%). Unfortunately, it was not reported whether the linolenic acid belonged to the ω-6 (γ-linolenic acid) or to the ω-3 series (α-linolenic acid) of fatty acids (Singh *et al.*, 1996a).

Along with the above given fatty acids a lipid soluble compound from a fraction isolated from the leaves of *O. gratissimum* grown in Nigeria has been reported. The compound was identified as a long chain unsaturated fatty acid, arachidonic acid (Onajobi 1986).

Angers *et al.* (1996 a) reviewed the existing literature and reported their own results on the fixed oil of basil seeds (Angers *et al.*, 1996 b). The oil content of the seeds of *Ocimum* species (*O. basilicum, O. canum*, *O. sanctum* and *O. gratissimum*) is on average 22% with a minimum of 18% (*O. canum*) and a maximum of 26% in the seeds of *O. basilicum*. The oils were rich in triacylglycerides (94–98%) and contained 1–3% of monoacylglycerides and diacylglycerides. α-Linolenic acid was the main compound in the seed oil of *O. basilicum* (57.4–62.5%) and *O. canum* (64.8%) and the major compound in the oils of *O. gratissimum* (47.4%) and *O. sanctum* (43.8%). The other compounds reported were linoleic acid (17.8–31.3%), oleic acid (8.5–13.3%) and palmitic acid (6.1–11.0%). The content of stearic acid varied from 2.0 to 4.0%.

Also small amounts of palmitoleic (0.2–0.3%) and γ-linolenic acid (0.1–0.3%) and traces of arachidic and eicosenic acids were found (Angers *et al.*, 1996b).

The older literature, reviewed by Angers *et al.* (1996 a), showed that instead of α-linolenic acid, a less unsaturated fatty acid, linoleic acid, has been found as the main compound of the fixed seed oil of *O. canum* (60.4%) (Patwardan 1930), in the seeds of *O. pilosum* (56.3%) (Khan *et al.*, 1961) and in the fixed oil of *O. sanctum* (66.1%) (Nadkarni and Patwardhan 1952). The content of α-linolenic acid was 21.3%, 18.5% and 15.7%, respectively. In these cases the content of stearic and oleic acid was approximately the same as described by Angers *et al.* (1996b).

The seed oil of *O. kilimandscharicum* contained α-linolenic acid as the main compound (65.0%). The oil also contained 16.2% of linoleic and 5.3% of oleic acid (Barker *et al.*, 1950). Almost the same result was reported by Henry and Grindley (1944). They found that the seed oil of *O. kilimandscharicum* contained 61% α-linolenic acid, 14% linoleic acid and 17% oleic acid. The seed oil of *O. viride* was almost as rich in α-linolenic and linoleic acid (39.2% and 32.5%, respectively) as the seed oil of *O. kilimandscharicum.* The content of oleic acid was 14.5% (Barker *et al.*, 1950).

The total of linoleic and linolenic acid in the fixed oil of *Ocimum* species appears to be fairly constant for all species. Their content ranges from 78% to 81%, except for *O. sanctum* where it is only 71%. The average content of unsaturated fatty acids, including α-linolenic, linoleic and oleic acid, is 89%. The most abundant saturated fatty acids are palmitic and stearic acid, ranging from 6.1% to 11.0% and from 2.0% to 4.0%, respectively. The composition of basil seed oil suggests that the oil would be suitable for industrial purposes, much in the same way as linseed oil (*Linum usitatissimum* L.). Linseed oil contains about 50–55% α-linolenic acid (Angers *et al.*, 1996).

FLAVONOIDS

So far only a few studies concerning the flavonoids in the genus *Ocimum* have been carried out. Two unusual flavones from *O. canum* growing in Somalia are reported. The flavones identified were navadensin (5,7-dihydroxy-6,8,4'-trimethoxyflavone) and salvigenin (5-hydroxy-6,7,4'-trimethoxyflavone). The amount of these flavones were 0.1% of the dry weight of the material that consisted of air-dried leaves and flowers of the plant (Xaasan *et al.*, 1980). The structures of these flavones are presented in Figure 3.1.

Basil herb (*O. basilicum*) contains apart from essential oil (0.5–1.5%) also flavonoid glycosides (0.6–1.1%) and flavonoid aglycones (Viorica 1987). A flavone, xanthomicrol (5,4'-dihydroxy-6,7,8-trimethoxyflavone) was isolated from the leaves of a Nigerian *O. basilicum* (Fatope and Takeda 1988). Three flavones, eriodictyol, eriodictyol-7-glucoside and vicenin-2 (apigenin di-C-glycoside), have been identified from the leaves of *O. basilicum* grown in Greece (Skaltsa and Philianos 1990). Three flavones have also been isolated from the leaves of *O. sanctum,* these being vicenin (apigenin-6,8-C-glycoside), galuteolin (luteolin-5-O-glycoside) and cirsilineol (5,4'-dihydroxy-6,7,3'-trimethoxyflavone) (Nörr and Wagner 1992).

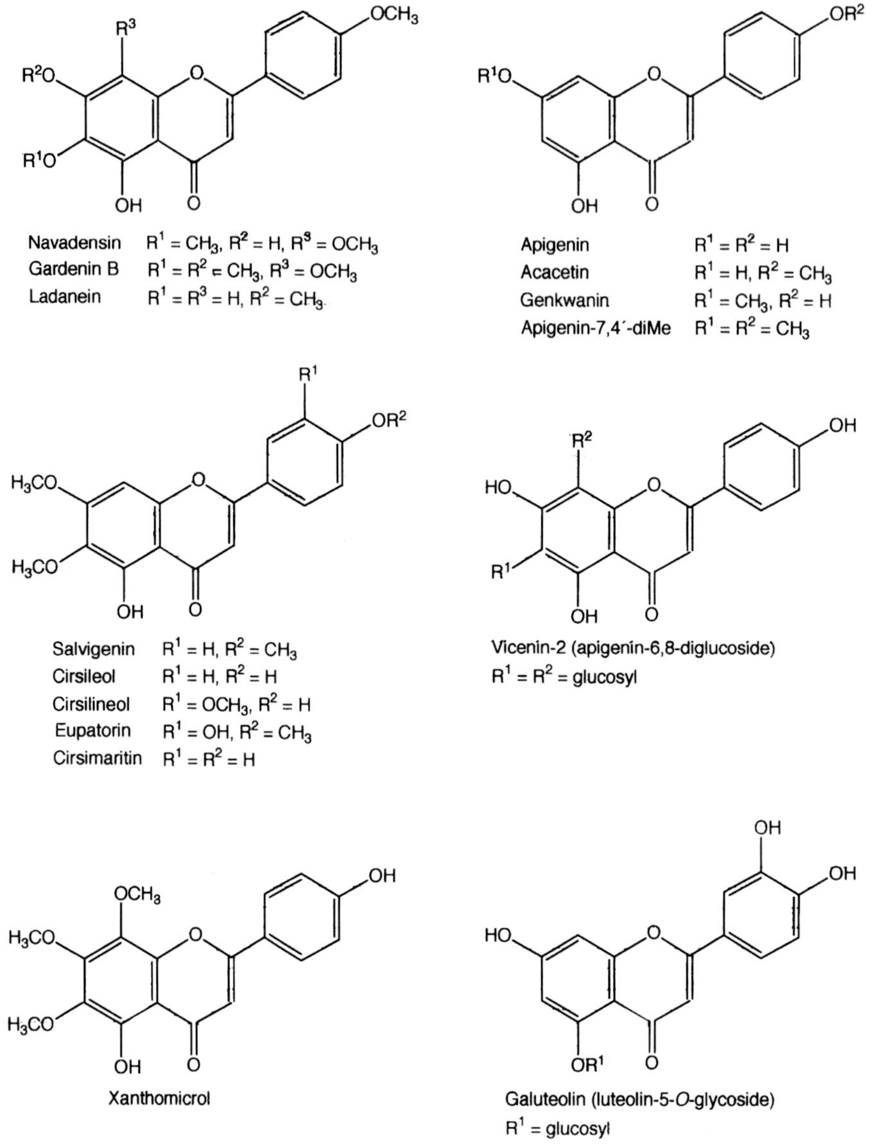

Navadensin	$R^1 = CH_3$, $R^2 = H$, $R^3 = OCH_3$
Gardenin B	$R^1 = R^2 = CH_3$, $R^3 = OCH_3$
Ladanein	$R^1 = R^3 = H$, $R^2 = CH_3$

Apigenin	$R^1 = R^2 = H$
Acacetin	$R^1 = H$, $R^2 = CH_3$
Genkwanin	$R^1 = CH_3$, $R^2 = H$
Apigenin-7,4′-diMe	$R^1 = R^2 = CH_3$

Salvigenin	$R^1 = H$, $R^2 = CH_3$
Cirsileol	$R^1 = H$, $R^2 = H$
Cirsilineol	$R^1 = OCH_3$, $R^2 = H$
Eupatorin	$R^1 = OH$, $R^2 = CH_3$
Cirsimaritin	$R^1 = R^2 = H$

Vicenin-2 (apigenin-6,8-diglucoside)
$R^1 = R^2$ = glucosyl

Xanthomicrol

Galuteolin (luteolin-5-*O*-glycoside)
R^1 = glucosyl

Figure 3.1 Flavonoids found in *Ocimum* species.

Navadensin and salvigenin that were found in the leaves of a Somalian *O. canum* were also identified from leaves of other *Ocimum* species (Grayer *et al.*, 1996). Grayer *et al.* examined external leaf flavonoids from 16 accessions of *O. basilicum* belonging to different cultivars and varieties. Accessions used for the chemical study were as follows: six accessions of *O. basilicum* var. *basilicum* (flat leaves) originating from Italy, the Netherlands, India, Yemen or the U.S.A, three accessions of *O. basilicum* var.

basilicum (convex leaves) originating from the U.S.A., one accession of an intermediate *O. basilicum* var. *basilicum*/var. *purpurascens* of Indian origin, three accessions of *O. basilicum* var. *purpurascens* one of which originated from Thailand, one from Israel, and one of unknown origin. Along with the above given species, also one accession of *O. basilicum* var. *difforme* originating from the U.K., two accessions of *O. basilicum* cv. Dark Opal from the U.K. and one accession of *O.* x *citriodorum* Vis. from Thailand and one accession of *O. minimum* originating from Turkey. Navadensin and salvigenin were found to be the major flavone aglycones. In addition 10 minor flavone aglycones were reported (Table 3.2). The flavones not listed in Table 3.2 are cirsileol that varied from trace amounts to 6.5%, apigenin (traces–3.8%), acacetin (traces–11.2%), genkwanin (traces–2.8%) and ladanein (traces–9.3%). Among the *Ocimum* species studied, the relative amounts of the total flavones varied in freeze-dried leaves from 3.15 to 59.53 calculated from the HPLC peak areas for a fixed sample amount. This shows that the total flavone concentration varies by a factor of 20 within *O. basilicum*. Grayer *et al.* (1996) did not find xanthomicrol in these species although it had been earlier reported in the leaves of a Nigerian *O. basilicum* (Fatope and Takeda 1988).

Lemberkovics *et al.* (1996) studied the essential oil, polyphenols and flavonoids of *O. basilicum* of Hungarian origin during the ontogeny. They found that the accumulation of flavonoid glycosides began in the budding stage and increased during flowering. The formation of flavone aglycones as well as tannins and polyphenols began just after sowing and increased during the vegetation period. The total flavonoid content varied from 0.28 to 0.61%. In this study two glycosides of quercetin, rutin and isoquercitrin, were found. These two flavonoids had been identified in *O. basilicum* already in 1993 by Nguyen *et al.* and Lemberkovics *et al.*

PHENOLIC ACIDS

In a search for the active principle of the leaves of *O. sanctum* a group of phenolic compounds was isolated. The isolated compounds were identified as rosmarinic acid,

Table 3.2 Percentage flavone composition of *Ocimum* leaves (Grayer *et al.*, 1996)

Taxon and Accession	*Navadensin*	*Salvigenin*	*Cirsilineol + Eupatorin*	*Apigenin 7,4′ -dimethyl Ether*	*Cirsimaritin*	*Gardenin B*
O. basilicum						
var. *basilicum* (9)	6.4–44.5	38.4–70.1	tr.–5.8	0.9–4.8	3.0–7.0	tr.–10.9
var. *basilicum*/var.						
purpurascens (1)	9.6	66.3	2.5	5.5	4.9	8.0
var. *purpurascens* (3)	5.6–32.7	38.8–60.7	3.6–6.5	2.6–4.7	1.7–3.8	tr.–15.9
var. *difforme* (1)	48.0	36.4	0.9	12.8	1.8	1.3
cv. Dark Opal (2)	51.7–53.0	28.0–37.0	5.0–5.5	2.1–6.5	2.9–4.2	tr.
O. x *citriodorum* (1)	38.7	42.0	3.3	3.7	3.4	3.6
O. minimum (1)	58.4	20.2	tr.	6.7	tr.	2.2

tr. = traces.

gallic acid, gallic acid methyl ester, gallic acid ethyl ester, protocatechuic acid, vanillic acid, 4-hydroxybenzoic acid, vanillin, 4-hydroxybenzaldehyde, caffeic acid and chlorogenic acid. In addition two phenylpropane glycosides, 4-allyl-1-O-β-D-gluco-pyranosyl-2-methoxybenzene (eugenyl-β-D-glucoside) and 4-allyl-1-O-β-D-glucopyranosyl-2-hydroxybenzene, were identified (Nörr and Wagner 1992). The structures of these phenolic acids are presented in Figure 3.2.

Nguyen *et al.* (1993) investigated the variation of the content of some secondary products during the vegetation period in a basil herb (*O. basilicum*) grown in Hungary. Among others, caffeic and rosmarinic acid were identified as the main components and were reported present in all the growing stages. The precursor of the cinnamates, *p*-coumaric acid, has been isolated and identified from the leaves of *O. basilicum* (Skaltsa and Philianos 1990). A phenolic acid ester, triacontanol ferulate (ca. 0.01%) was identified from the stem bark of *O. sanctum* (Sukari *et al.*, 1995).

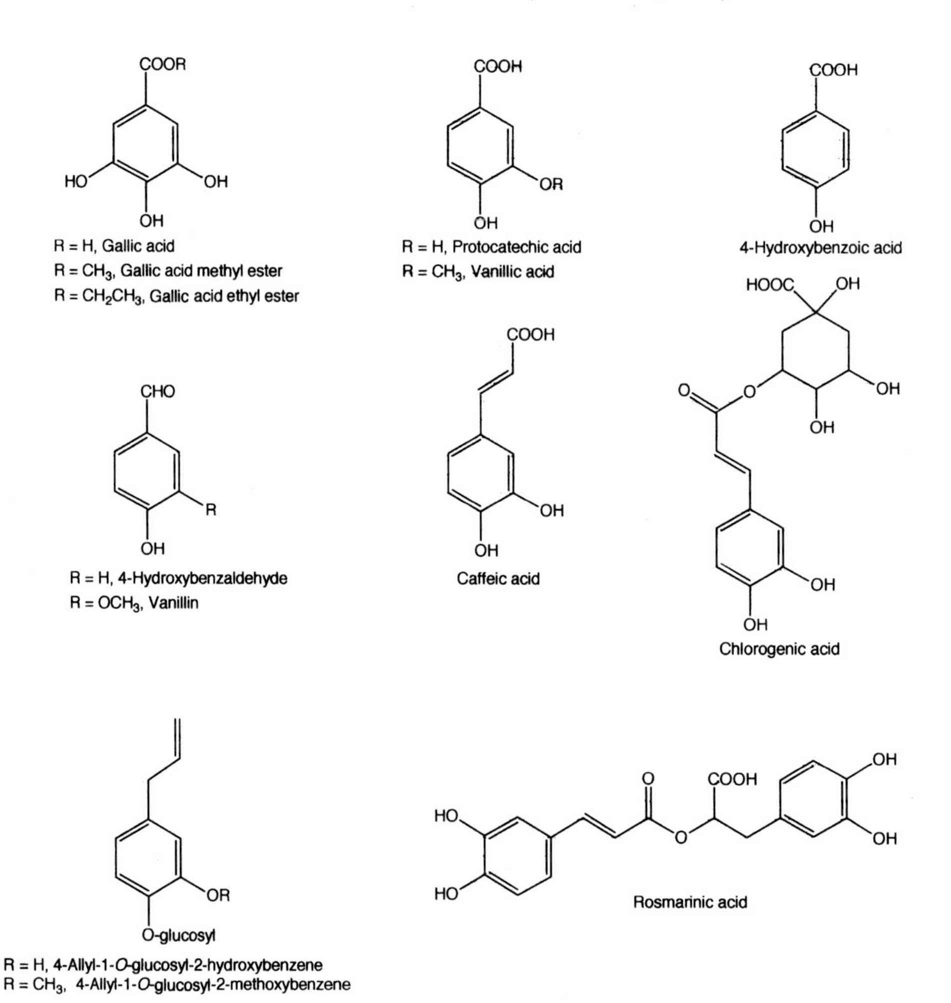

Figure 3.2 Phenolic acids found in *Ocimum* species.

TRITERPENES AND STEROIDS

A mixture of ursolic and oleanolic acid (0.2% d.w.) was isolated from the leaves and flowers of a Somalian *O. canum* (Xaasan *et al.*, 1980). These two triterpenes have been reported also in *O. sanctum* (Balanehru and Nagarajan 1994). β-Sitosterol have been detected in basil herb (*O. basilicum*) (List and Hörhammer 1977, Viorica 1987). Sukari *et al.* (1995) identified it in mixture of stigmasterol and β-sitosterol (0.02%, d.w.) also from the stem bark of *O. sanctum*. The structures of ursolic, oleanolic acid and β-sitosterol are shown in Figure 3.3.

OTHERS

Basil herb (*O. basilicum*) contains apart from essential oil and flavonoids also tannins and polyphenols (2.2–2.3%) (Viorica 1987). Nguyen *et al.* (1993) and Lemberkovics *et al.* (1996) examined the accumulation of polyphenols and tannins during the vegetation period in *O. basilicum* and *O. sanctum* herbs of Hungarian origin. The content of tannins varied during the vegetation period from about 1% to 5% in *O. basilicum* herb

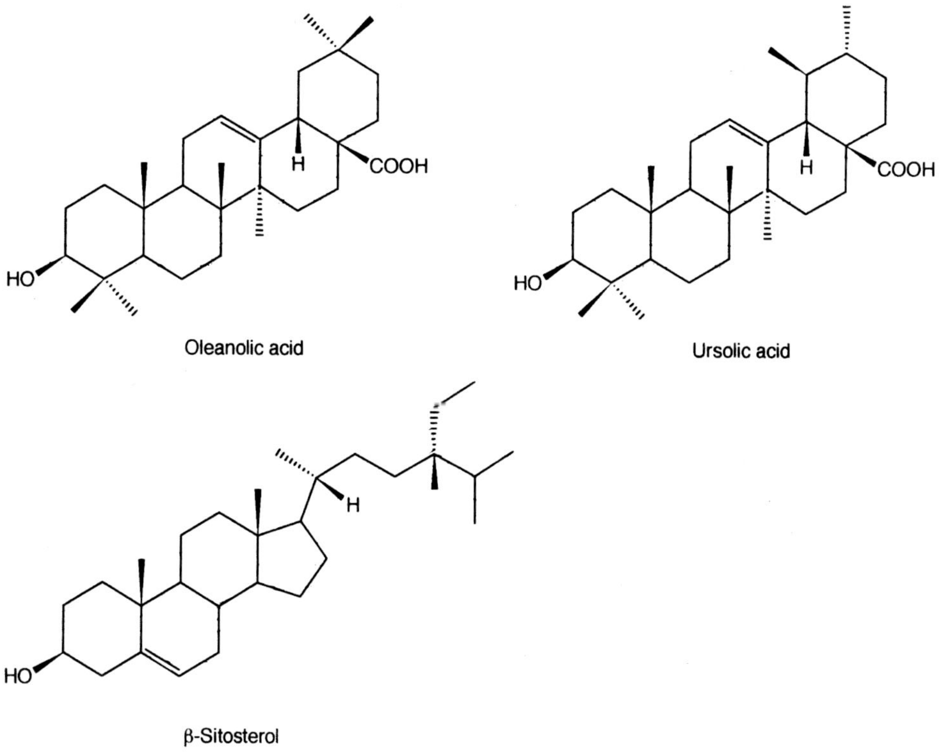

Oleanolic acid

Ursolic acid

β-Sitosterol

Figure 3.3 Triterpenes and a steroid (β-sitosterol) found in *Ocimum* species.

and in the stems, leaves and flowers of *O. sanctum* the tannin content varied from about 1% to 3.5% . The content of polyphenols in the leaves, stems and flowers of *O. basilicum* ranged during the same period from 2% to about 10% . The polyphenol content of *O. sanctum* was 2–5% (Nguyen *et al.*, 1993).

The content of polyphenols (5.0–8.5%) and tannins (2.6–4.1%) did not change much during early, full and late flowering. However, it decreased a little bit during the early productive and ripening stages (Lemberkovics *et al.*, 1996).

Some additional individual compounds from *Ocimum* species have been isolated, identified and discussed in the literature. Among them e.g. hydroquinone and two tentatively identified compounds (molecular formula either $C_{21}H_{37}O_4$ or $C_{21}H_{35}N_3O_3$ as deduced from computer analysis) isolated from *O. gratissimum* (Onajobi 1986), elsholtzione (2-isovaleryl-3-methylfuran isolated from *O. basilicum* var. *hispidum* (Lam.) Chiov. by Ruberto *et al.*, 1991) and hemicellulosic polysaccharides determined by Figuiredo-Ribeiro *et al.* (1992) from *O. nudicaule* Benth. var. *anisfolia* Giul. and a coumarin, aesculetin (Skaltsa and Philianos 1990).

REFERENCES

Angers, P., Morales, M.R. and Simon, J.E. (1996 a) Basil seed oils. In J. Janick (ed.), *Progress in New Crops*, ASHS Press, Arlington, USA, pp. 598–601.

Angers, P., Morales, M.R. and Simon, J.E. (1996 b) Fatty acid variation in seed oil among *Ocimum* species. *JAOCS*, **73**, 393–395.

Balanehru, S. and Nagarajan, B. (1994) Triterpenes intervene in adriamycin-induced histamine release in rodents. *Med. Sci. Res.*, **22**, 357–359.

Barker, C., Dunn, H.C. and Hilditch, T.P. (1950) African drying oils. V. Some Nigerian and Sudanese drying oils. *J. Soc. Chem. Ind.*, **69**, 71–75.

Domokos, J., Peredi, J., Palevitch, D. and Pitievsky, E. (1993) Studies of seed oils of basil (*Ocimum basilicum* L.) and summer savory (*Satureja hortensis* L.). *Acta Horticulture*, **344**, 312–314.

Fatope, M.O. and Takeda, Y. (1988) The constituents of the leaves of *Ocimum basilicum. Planta Medica*, **54**, 190.

Figueiredo-Ribeiro, C.L., Isejima, E.M., Dietrich, S.M.C. and Correa, J.B.C. (1992) Hemicellulosic polysaccharides from the xylopodium of *Ocimum nudicaule*: Changes in composition in dormancy and sprouting. *Annals of Botany*, **70**, 405–408.

Grayer, R.J., Bryan, S.E., Veitch, N.C., Goldstone, F.J., Paton, A. and Wollenweber, E. (1996) External flavones in sweet basil, *Ocimum basilicum*, and related taxa. *Phytochemistry*, **43**, 1041–1047.

Henry, A.J. and Grindley, D.N. (1944) The oils of the seeds of *Ocimum kilimandscharicum*, *Euphorbia calysina*, *E. erythraeae*, *Sterculia tomentosa* and *Trichilia emetica. J. Soc. Chem. Ind.*, **63**, 188–190.

Khan, S.A., Qureshi, M.I., Bhatty, M.K. and Karimulah (1961) Composition of the seed oils of *Salvia spinosa* and *Ocimum pilosum* (*O. basilicum*). *Pakistan J. Sci. Res.*, **13**, 41–43.

Lemberkovics, É., Petri, G., Nguyen, H. and Máthé, I. (1996) Relationships between essential oil and flavonoid biosynthesis in sweet basil. In L.E. Craker, L. Nolan, and K. Shetty (eds.), *Proceedings Int. Symp. Medicinal and Aromatic Plants*, *Acta Hort.*, **426**, pp. 647–655.

Lemberkovics, É., Nguyen, H., Máthé, I., Tarr, K., Petri, G. and Vitányi, G. (1993) Formation of essential oil and phenolic compounds during the vegetation period in *Ocimum basilicum. Planta Medica*, **59**, Suppl., A700–A701.

Leung, A.Y. and Foster, S. (1996) *Encyclopedia of Common Natural Ingredients used in foods, drugs, and cosmetics*, 2nd ed., John Wiley & Sons, New York, USA.

List, P.H. and Hörhammer, L. (1977) Hagers Handbuch der Pharmazeutischen Praxis, 4th ed., Band VI A, Springer Verlag, Berlin-Heidelberg, Germany.

Malik, M.S., Rafique, M., Sattar, A. and Khan, S.A. (1987) The fatty acids of indigenous resources for possible industrial applications. Part XII: The fatty acid composition of the fixed oils of *Ocimum sanctum* and *Salvia aegyptica* seeds. *Pakistan J. Sci. Ind. Res.*, **30**, 369–371.

Malik, M.S., Sattar, A. and Khan, S.A. (1989) The fatty acids of indigenous resources from possible industrial applications. Part XVII: The fatty acid composition of the fixed oils of *Ocimum basilicum* and *Ocimum album* seeds. *Pakistan J. Sci. Ind. Res.*, **32**, 207–208.

Nadkarni, G.B. and Patwardan, V.A. (1952) Fatty oil from the seeds of *Ocimum sanctum* Linn. (Tulsi). *Current Sci.*, **21**, 68–69.

Nguyen, H., Lemberkovics, É., Tarr, K., Máthé Jr., I. and Petri, G. (1993a) A comparative study on formation of flavonoid, tannin, and polyphenol contents in ontogenesis of *Ocimum basilicum* L. Part I. *Acta Agronomica Hungarica*, **42**, 31–39.

Nguyen, H., Lemberkovics, É., Tarr, K., Máthé Jr., I. and Petri, G. (1993b) A comparative study on formation of flavonoid, tannin, and polyphenol contents in ontogenesis of *Ocimum basilicum* L. Part II. *Acta Agronomica Hungarica*, **42**, 41–50.

Nörr, H. and Wagner, H. (1992) New constituents from *Ocimum sanctum. Planta Medica*, **58**, 574.

Onajobi, F.D. (1986) Smooth muscle contracting lipid-soluble principles in chromatographic fractions of *Ocimum gratissimum. Journal of Ethnopharmacology*, **18**, 3–11.

Patwardan, V.A. (1930) Thesis, University of Bombay. In T.P. Hildtich (ed.), *The Chemical Composition of Natural Fats*, John Wiley, New York, 1940.

Ruberto, G., Spadaro, A., Piattelli, M., Piozzi, F. and Passannanti, S. (1991) Volatile flavour components of *Ocimum basilicum* var. *hispidum* (Lam.) Chiov. *Flav. Fragr. J.*, **6**, 225–227.

Singh, S., Majumdar, D.K. and Yadav, M.R. (1996) Chemical and pharmacological studies on fixed oil of *Ocimum sanctum. Indian Journal of Experimental Biology*, **34**, 1212–1215.

Singh, S., Majumdar, D.K. and Rehan, H.M.S. (1996) Evaluation of anti-inflammatory potential of fixed oil of *Ocimum sanctum* (Holybasil) and its possible mechanism of action. *Journal of Ethnopharmacology*, **54**, 19–26.

Skaltsa, H. and Philianos, S. (1990) Contribution á l étud chimique *d'Ocimum basilicum* L. *Plantes médicinales et phytothérapie*, **XXIV**, 193–196.

Sukari, M.A., Rahmani, M. and Lee, G.B. (1995) Constituents of stem barks of *Ocimum sanctum. Fitoterapia*, **LXVI**, 552–553.

Viorica, H. (1987) Polyphenols of *Ocimum basilicum* L. *Chujul Med.*, **60**, 340–344.

Xaasan, C.C., Ciilmi, C.X., Faarax, M.X., Passannanti, S., Piozzi, F. and Paternostro, M. (1980) Unusual flavones from *O. canum* . *Phytochemistry*, **19**, 2229–2230.

4. ESSENTIAL OIL OF *OCIMUM*

RAIMO HILTUNEN and YVONNE HOLM

Department of Pharmacy, P.O Box 56,
FIN-00014 University of Helsinki,
Finland

INTRODUCTION

The genus *Ocimum* belongs to the Lamiaceae family and consists of 50 to 150 species of herbs and shrubs. Plants of the genus *Ocimum* are collectively called basil (Simon *et al.*, 1990). According to Hegnauer (1966) the genus *Ocimum* comprises 50–60 species. In 1983 Czikow and Laptiew reported that there exist 150 species in the genus *Ocimum* and one year later Ruminska reported that about 60 species have been classified as belonging to this genus (Suchorska and Osinska 1992).

Basil oil is one of the most widespread oils. Currently the oil production is 15 tons in India, 7 tons in Bulgaria, 5 tons in Egypt, 4.5 tons both in Pakistan and the Comores, 2 tons in Israel and smaller amounts in Yugoslavia (1 ton), USA (1 ton), Madagascar (1 ton), Réunion (0.5 ton), Albania (0.5 ton), Hungary (0.3 ton) and in Argentina (0.2 ton) (Lawrence 1992).

The market for basil oil is dominated by European and Egyptian basil oil. The European or sweet basil is cultivated and the oil produced in the Mediterranean region and in the United States. These two types of oil are considered to be of the highest quality and the ones which produce the finest odour. Sweet basil oil is characterized by high concentrations of linalool (30–90%) and methyl chavicol (50–90%). The other characteristic oil compounds are 1,8-cineole, eugenol and methyl eugenol. In addition to oxygenated monoterpenes and phenylpropanoids, also small amounts or at least traces of monoterpene and sesquiterpene hydrocarbons are typically present in both oils.

Reunion basil oil, also known as Comoro, African or exotic basil oil, is another type of basil oil traded on the international market. This type of basil is cultivated and the oil produced on Reunion, Madagascar, in many parts of Africa and occasionally on the Seychelles. Reunion basil oil is rich in methyl chavicol with a high content of camphor but low contents of linalool, 1,8-cineole and eugenol.

Basil is known to occur as several chemotypes or cultivars which differ in their essential oil composition. The variation in chemical composition of basil oils is thought to be mainly due to polymorphism in *Ocimum basilicum* L., which is caused by interspecific hybridization (Hasegawa *et al.*, 1997). The morphological diversity within basil species has been accentuated by centuries of cultivation. Great variation in pigmentation, leaf shape and size and pubescence are seen (Simon *et al.*, 1990).

ESSENTIAL OIL OF BASIL

The essential oils of basil are generally obtained by steam distillation or hydrodistillation from flowering tops (French sweet basil oils) and leaves of basil plants. The oil yield generally ranges from 0.2 to 1.0 but can be as high as 1.7%, depending on the source and phenological stage of the plants. Also whole fresh flowering herbs (American sweet basil oils) have been used as the starting material for the commercial production of volatile oils. Supercritical fluid extraction (SFE) has also been applied to the isolation of volatiles from basil (Manninen *et al.*, 1990, Reverchon 1994, Pluhar *et al.*, 1996, Lachowicz *et al.*, 1997).

Volatiles in the essential oils of *Ocimum* species are mainly derived either from the phenylpropanoid or mevalonic acid metabolism. Biogenetic pathways and the formation of volatiles, such as terpenes from the isoprenoid metabolism and phenylpropanoids from the cinnamic acid metabolism, have been studied and discussed intensively during the last 30–40 years (Schreier 1984). The biosynthesis of the essential oil compounds of *Ocimum* species will not be discussed in this context.

CHEMICAL COMPOSITION

The chemical composition of essential oils of plants in the genus *Ocimum,* especially *O. basilicum,* have been the object of many studies since the 1930's. The older chemical data of *Ocimum* oils is summarized by Günther (1949), Gildemeister and Hoffmann (1961), Hegnauer (1966) and Hoppe (1975). Especially Lawrence (1971, 1972, 1978, 1980, 1982, 1986a, 1986b, 1987, 1988, 1989, 1992a, 1992b, 1992c, 1995, 1997) has done a great work in summarizing and reporting the results of the chemical composition of basil oils published during the last 25 years. Also the studies made by Sobti *et al.* (1976), Gulati (1977), Gulati *et al.* (1977a, 1977b), Peter and Rémy (1978), Vernin *et al.* (1984), Nykänen (1987), Ekundayo *et al.* (1989), Gaydou *et al.* (1989), Fun and Baerheim Svendsen (1990) and particularly Zola and Garnero 1973, Nykänen 1987, Chien 1988, Lawrence 1988, Hasegawa *et al.* (1997) have increased our knowledge about the chemical composition of basil oils. Nowadays approximately 140 components of basil oil (*O. basilicum*) are known, including more than 30 monoterpenes, almost 30 sesquiterpenes, about 20 carboxylic acids, 11 aliphatic aldehydes, 6 aliphatic alcohols, about 20 aromatic compounds and about 20 compounds belonging to other groups than those mentioned above (Table 4.1).

In general basil oils are characterized by oxygenated monoterpenes (Table 4.2) and phenylpropane derivatives (Tables 4.3–4.9). In addition to oxygenated monoterpenes almost all of the known monoterpene hydrocarbons are found in basil oils (Table 4.1). Usually their concentrations are marginal except for the ocimenes, γ-terpinene and *p*-cymene in the oil of *O. gratissimum.*

An oil of *O. basilicum* var. *hispidum* had an exceptional chemical composition containing an acyclic monoterpene ketone, dihydrotagetone, as its main compound (82.3%) and smaller amounts of compounds structurally related to it, including ipsenone (0.2%), *cis*-tagetone (0.5%) and *trans*-tagetone (0.2%) (Ruberto *et al.*, 1991).

Table 4.1 Chemical constituents of basil oils

Chemical constituents
Monoterpene Hydrocarbons
3-carene
p-cymene
limonene
myrcene
cis-β-ocimene
trans-β-ocimene
cis-allo-ocimene
trans-allo-ocimene
α-phellandrene
β-phellandrene
α-pinene
β-pinene
sabinene
α-terpinene
γ-terpinene
terpinolene
α-thujene
Oxygenated Monoterpenes
borneol
bornyl acetate
camphor
1,8-cineole
citronellal
citronellol
citronellyl acetate
p-cymen-8-ol
fenchone
fenchyl acetate
α-fenchyl acetate
fenchyl alcohol
geranial
geraniol
geranyl acetate
isobornyl acetate
linalool
linalool oxide
cis-linalool oxide
trans-linalool oxide
linalool oxide (furanoid)
linalool oxide (pyranoid)
linalyl acetate
menthol
menthone
myrtenal
neral
nerol
neryl actate
trans-ocimene oxide
perilla aldehyde
terpinen-4-ol
α-terpineol
α-terpinyl acetate
α-thujone
β-thujone
cis-sabinene hydrate
trans-sabinene hydrate
Sesquiterpene Hydrocarbons
α-amorphene
cis-α-bergamotene
trans-α-bergamotene
bicyclogermacrene
bicycloelemene
α-bisabolene
β-bisabolene
β-bourbonene
ε-bulgarene
α-cadinene
δ-cadinene
γ-cadinene
calamenene
β-caryophyllene
α-caryophyllene
β-cedrene
2-*epi*-α-cedrene
α-copaene
β-copaene
α-cubebene
β-cubebene
cyclosativene
β-elemene
δ-elemene
γ-elemene
(elixene)
α-farnesene
E-β-farnesene
germacrene-D
germacrene-B
α-guaiene
δ-guaiene
β-gurjunene
γ-gurjunene
β-himachalene
α-humulene
isocaryophyllene
ledene
α-muurolene
β-patchoulene
α-santalene
α-selinene
β-selinene
β-sesquiphellandrene
viridifloral
Oxygenated Sesquiterpenes
α-bisabolol
β-bisabolol
bulnesol
T-cadinol
10-*epi*-α-cadinol
trans-cadinol
β-caryophyllene oxide
cedrol
cubenol
elemol
β-eudesmol
γ-eudesmol
farnesol
nerolidol
ledol
spathulenol
Others
p-methoxy acetophenone
iso-amyl alcohol
benzene
butanal
β-damascenone
β-damascone
2,4-decadienal
decanol
dodecanol
δ-dodecalide

Table 4.1 (Continued)

cumin aldehyde	3-octanal	benzyl formiate
ethyl acetate	2,4-octadienal	benzyl benzoate
2-ethyl furan	3-octanone	*p*-methoxybenzaldehyde
2-methyl furan	octanol	chavicol
ethyl-2-methyl butyrate	3-octanol	methyl chavicol (= estragole)
methyl-2-methyl butyrate	*trans*- 2-octen-1-al	cinnamyl acetate
furfural	1-octen-3-ol	(*E*)-methyl cinnamate
heptanal	octyl acetate	(*Z*)-methyl cinnamate
hexanol	1-octen-3-yl acetate	*p*-methoxycinnamyl-
cis-hex-3-en-1-ol	pentanal	alcohol
5-methyl-2-heptanone	2-methyl-3-methoxy	coumarin
6-methyl-5-hepten-2-one	pyrazine	methyl eugenol
6-methyl-3-heptanone	tetra-methyl pyrazine	*p*-methoxy-cinnamaldehyde
cis-3-hexenol	quinoline	eugenol
3-hexenyl acetate	undecylaldehyde	acetyl eugenol
cis-3-hexenyl acetate	*Aromatic Compounds*	isoeugenol
cis-3-hexenyl benzoate		phenyl ethyl alcohol
methyl isovalerate	*cis*-anethole	phenyl ethyl acetate
methyl jasmonate	*trans*-anethole	α-p-dimethyl styrene
methyl epi-jasmonate	anisaldehyde	methyl salicylate
cis-jasmone	benzaldehyde	methyl thymol
trans-jasmone	benzyl alcohol	thymol
dibutyl octanediotate	benzyl acetate	vanillin

Chemical compounds in essential oils are found either as free terpenoids and phenylpropanoids or bound to sugar moieties. Lang and Hörster (1977) studied the production and accumulation of essential oil in *O. basilicum* callus and suspension cultures. They found that both free and sugar bound monoterpenoid and phenylpropanoid substances existed in morphologically undifferentiated callus-suspension cultures. However, the sugar bound form seemed to be the preferred accumulation form of the studied substances.

MAIN AND MAJOR COMPOUNDS IN *OCIMUM* OILS

In the following the main and some major compounds detected in the essential oils of *Ocimum* are discussed. In this connection the main compounds are those which represent 50% or more of the total oil. However, a relative percentage of 50% does not mean that the absolute amount of the compound is 50% of the total oil. The percentage composition of an essential oil is usually based on gas chromatographic analysis and on the assumption that the detector response is the same for all the involved compounds. This assumption is, of course, false. The major compounds are all the constituents, whose relative percentages are somewhere between 20–50% of the total oil. *Ocimum* oils generally consist of one or two, seldomly more than four

Table 4.2 Predominant oxygenated monoterpenes in basil oils

Constituent	*Ocimum spp.*	*References* [1]
linalool	*O. basilicum*	see Table 3
camphor	*O. canum*	Hegnauer 1966, Ntezurubanza *et al.*, 1986, Xaasan and Cabdulraxmaan 1981
	O. basilicum var. *glabratum*	Gupta 1994
	O. kilimandscharicum	Prasad *et al.*, 1986
1,8-cineole	*O. canum*	
	O. kilimandscharicum	Ntezurubanza *et al.*, 1984
	O. micranthum	Charles *et al.*, 1990
	O. keniense	Mwangi *et al.*, 1994
	O. basilicum	Holm *et al.*, 1989
	O. gratissimum [2]	Cheng and Liu 1983
geraniol	*O. canum*	Sobti and Pushpangadan 1982
	O. basilicum	Sobti and Pushpangadan 1982
citral	*O. citriodorium*	Simon *et al.*, 1990
neral	*O. citriodorium*	Grayer *et al.*, 1996
geranial	*O. x citriodorium* [3]	Grayer *et al.*, 1996
thymol	*O. gratissimum*	El-Said *et al.*, 1969, Sobti 1979, Janssen *et al.*, 1989, Thomas 1989, Hegnauer 1966, Gildemeister and Hoffmann 1961, Sofowora 1970, Sainsbury and Sofowora 1971, Ntezurubanza *et al.*, 1987, Pino *et al.*, 1996
	O. basilicum	Fatope and Takeda 1988
	O. viride	Ekundayo 1986

[1] see also Guenther 1949, Gildemeister and Hoffmann 1961.
[2] 1,8-cineole as the major compound in the stem oil.
[3] a hybrid between *O. basilicum* and *O. americanum.*

compounds, with a proportional share of the total oil greater than 20%. Therefore, also smaller compounds (5–10%) can be considered to be major compounds.

Monoterpene Hydrocarbons

The amounts of monoterpene hydrocarbons in the essential oils of *Ocimum* species are generally very low ranging from traces to only a few percent only. The average percentage is less than one percent of the total oil although there are some exceptions to this.

Limonene and myrcene

The concentration of limonene and myrcene is very low in the essential oils of the *Ocimum* species as is also the proportional amount of most monoterpene hydrocarbons. The content of limonene varies from 0 to 1.0% and myrcene from 0 to 1.2% (*e.g.* Lawrence 1978, 1980, 1982, 1988, 1989, 1992, 1995, 1997, Chien 1988, Tapanes

1985). Only a few cases are known where the amounts of limonene and myrcene exceed the levels given above. An essential oil of *O. canum* of Somalian origin contained as much as 13% limonene (Xaasan *et al.*, 1981). Chien (1988) and Lawrence (1988) reported limonene contents of 3.6% and 2.1%, respectively, in *O. basilicum* oil. Conan (1977) made a comparative study on the chemical composition of basil oils originating from South Africa, the Comoros, France and Egypt. Samples from Egypt contained 4.7–9.3% limonene, and the rest of the samples 2.0–4.9% limonene. The myrcene content varied from 2.9 to 8.2% in an essential oil of *O. gratissimum* grown in Cuba (Pino *et al.*, 1996). On average the content of myrcene in *O. basilicum* oil is about 0.5%.

p-Cymene and γ-terpinene

The oil of *O. gratissimum* is rich in γ-terpinene and *p*-cymene. Dro (1974) examined the chemical composition of an *O. gratissimum* essential oil from plants grown in Nigeria. The amount of *p*-cymene varied from 10.3 to 32.4% and γ-terpinene together with *trans*-β-ocimene from 10.1 to 27.0%. Another Nigerian *O. gratissimum* oil contained 16.2% *p*-cymene (Sainsbury and Sofowora 1971). Also *O. gratissimum* grown in Rwanda have been found to be rich in γ-terpinene (3.4–22.9%) and *p*-cymene (5.8–18.3%) (Ntezurubanza *et al.*, 1987). Also *O. gratissimum* grown in Cuba contained high levels of γ-terpinene (7.0–8.2%) and *p*-cymene (12.8–14.0%) (Pino *et al.*, 1996). The essential oil of *O. gratissimum* of Brazilian origin was very rich in *p*-cymene, containing as much as 29.7% (Maia *et al.*, 1988). The essential oil of *O. gratissimum* from Madagascar contained 13.2% γ-terpinene (De Medici *et al.*, 1992).

In general the content of *p*-cymene and γ-terpinene in the essential oil of the other *Ocimum* species is very low. *p*-Cymene ranges from 0 to 0.5% and γ-terpinene from 0 to 0.2% in *O. basilicum* and from 0 to 0.6% in *O. gratissimum* (*e.g.* Lawrence 1978, 1986, 1987, 1992, 1995, 1997).

Ocimenes

Both (*Z*)-β-ocimene and (*E*)-β-ocimene have been identified in the essential oil of *Ocimum.* Generally the proportional shares of these compounds are small in the total oil of *O. basilicum*, *O. sanctum* and *O. gratissimum*, varying from trace amounts to less than 4% (*e.g.* Lawrence 1978, 1980, 1986, 1987, 1989, 1992, 1995, Cheng and Liu 1983, Perez-Alonso *et al.*, 1995, Laakso *et al.*, 1990). However, high concentrations of both (*Z*)-β-ocimene and (*E*)-β-ocimene have been reported in the oil of *O. gratissimum.* Zamuurenko *et al.* (1986) found that a Russian *O. gratissimum* oil contained 0.3–14.5% (*Z*)-β-ocimene and trace amounts to 12.1% of (*E*)-β-ocimene. Yu and Cheng (1986) found 12.9% β-ocimene and 0.3% α-ocimene in a Chinese *O. gratissimum* oil. Lawrence (1997) presumed that these compounds were incomplete characterizations of (*Z*)-β-ocimene and (*E*)-β-ocimene, respectively.

Higher concentrations of ocimenes have also been detected in the essential oil of *O. trichodon.* An essential oil of *O. trichodon* grown in Rwanda contained 2.2–15.9% (*Z*)-β-ocimene and 1.4–8.8% (*E*)-β-ocimene (Ntezurubanza *et al.*, 1986). Also the essential oil of *O. urticifolium* (syn. *O. suave*) from Rwanda was rich in ocimenes and particularly rich in (*Z*)-β-ocimene, which was a major compound (41.0%) in the oil.

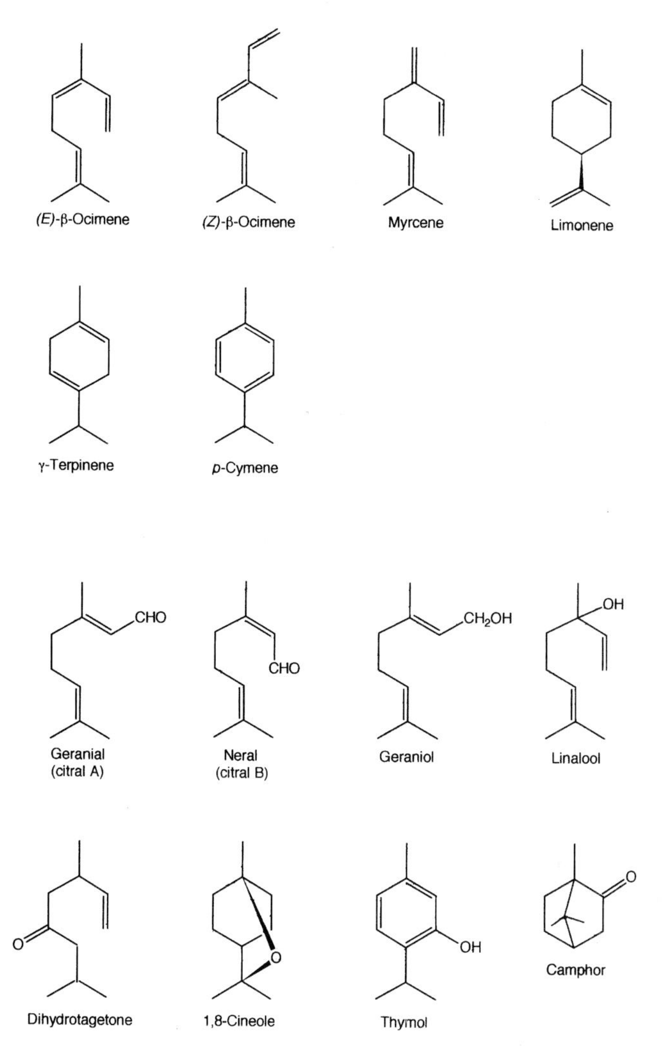

Figure 4.1 Monoterpene hydrocarbons and oxygenated monoterpenes found in the essential oils of *Ocimum* species.

In this oil the amount of (*Z*)-β-ocimene varied from 12.6 to 41.0% and the amount of (*E*)-β-ocimene was 13.5% (Janssen *et al.*, 1989).

Oxygenated Monoterpenes as Main Constituents

Linalool

Linalool is one of the most common compounds in basil oil. It is characteristic of the essential oil of European or sweet basil (*O. basilicum*), but it is also common in oils of *O. canum* (*e.g.* Günther 1949, Gulati *et al.*, 1977b, Sobti *et al.*, 1976, Ntezurubanza *et al.*, 1985, Gupta and Sobti 1993, Gupta and Tawa 1997), in *O. menthaefolium* and in

O. sanctum (Hegnauer 1966). Its percentage proportion in the oils ranges from traces up to almost 90%. Linalool may co-exist with methyl chavicol, eugenol, methyl cinnamate, 1,8-cineole and with some other monoterpenes such as geraniol, citronellol, camphor, *p*-cymene, myrcene, and many others. Table 4.3 depicts the origin of the studied basil plants, the content of linalool and references in which linalool has been detected as the main compound or at least one of the major constituents.

Camphor

Hegnauer (1966) reported a camphor type essential oil of *O. basilicum* originating from Reunion, the Comoros and the Seychelles. Sobti and Pushpangadan (1982) reported that the camphor chemotype of *O. basilicum* contains 10–15% of camphor. Otherwise the reports of the chemical composition of *O. basilicum* essential oils, published during the last 20–30 years, do not include oils rich in camphor. For instance Fleisher and Fleisher (1992) found only 1.3% of camphor in the oil of *O. basilicum* grown in Israel.

Also the oils of *O. gratissimum* grown in Russia (Zamureenko *et al.*, 1986) and in India (Khanna *et al.*, 1988) have low levels of camphor (traces–6.7%). However, the oils of some *O. gratissimum* varieties such as *O. gratissimum* var. *glabratum* (Gupta 1994), *O. gratissimum* var. *camphorata* (Hegnauer 1966) and *O. gratissimum* var. *intermedia* (Gildemeister and Hoffmann 1961), as well as the oil of *O. canum* (Hegnauer 1966, Hoppe 1975, Xaasan *et al.*, 1981) and *O. kilimandscharicum* (Hegnauer 1966, Gildemeister and Hoffmann 1961) have been reported to be rich in camphor. The camphor content of these oils varied from 23 to 70%, being generally 50 to 70% in the oils of *O. kilimandscharicum* and *O. canum*.

1,8-Cineole

1,8-Cineole is one of the key compounds in the classification of the chemical composition of essential oils of the *Ocimum* species. Among *Ocimum* oils the amount of 1,8-cineole varies from traces to more than 60%. In the oil of *O. kilimandscharicum* and *O. keniense* 1,8-cineole was the main compound. Ntezurubanza *et al.* (1984) reported a new chemotype, 1,8-cineole type, of the essential oil of *O. kilimandscharicum* grown in Rwanda. This oil contained 62.2% of 1,8-cineole. 1,8-Cineole (38.4%) was reported as the main compound in the essential oil of *O. keniense* collected in Kenya (Mwangi *et al.*, 1994).

The oil of *O. gratissimum* is either rich or poor in 1,8-cineole. Cheng and Liu (1983) examined an essential oil of *O. gratissimum* originating from Taiwan and found that the stem oil contained 40.2% and the flower oil 23.0% 1,8-cineole. In the leaf oil the percentage of 1,8-cineole was only 2.9%. Also an oil of *O. gratissimum* from Madagascar was rather rich in 1,8-cineole (12.0%) (De Medici *et al.*, 1992). Lower levels of 1,8-cineole in *O. gratissimum* essential oils have been reported by Ntezurubanza *et al.* (1987), Vostrowsky (1990) and Fun and Baerheim Svendsen (1990). The oils obtained from plants grown in Rwanda contained 0.3–3.5% 1,8-cineole (Ntezurubanza *et al.*, 1987), from the Amazon 3.3% (Vostrowsky 1990) and from Aruba 6.2% (Fun and Baerheim Svendsen 1990).

Table 4.3 Linalool as a predominant constituent in the essential oils of basil

Origin	*Linalool Content (%)*	*References*
Calabria	40.0	Günther 1949
India	45	
Egypt	not mentioned	Hörhammer *et al.*, 1964
Bulgaria	50–63	Georgiev and Genov 1973
France	39.1	Zola and Garnero 1973
Italy	43.8	Zola and Garnero 1973
Morocco	41.9	Zola and Garnero 1973
Egypt	38	Karawya *et al.*, 1974
Bulgaria	not mentioned	Hoppe 1975
	80.7–87.3	Skrubis and Markakis 1976
	not mentioned	Lang and Hörster 1977
South Africa, Egypt, India, Yugoslavia, France, Israel	66–86	Peter and Rémy 1978
Israel	37.5–55.4	Fleischer 1981
	30–35	Sobti and Pushpangadan 1982
Bulgaria, Egypt, France, Yugoslavia	56.0–70.0	Vernin *et al.*, 1984
Austria	40.7–75.5	Kartnig and Simon 1986
	62.3–62.4	Srinivas 1986
commercial oil (group 2)	43.8–52.0	Srinivas 1986
commercial oil (group 3)	41.9–54.4	Srinivas 1986
experimental oil (group 2)	23.0–75.4	Srinivas 1986
experimental oil (group 3)	63.8–74.2	Srinivas 1986
Nigeria	30.1	Ekundayo *et al.*, 1987
Hungary	33.8–46.7	Hälvä 1987
U.S.A.	7–62	Simon *et al.*, 1990
Portugal	32.2	Carmo *et al.*, 1989
Italy	46.0–50.0	Tateo *et al.*, 1989
Italy	45.7–63.1	Mariani *et al.*, 1991
Poland	54.1	Kostrzewa and Karwowska 1991
		Bernreuter and Schreier 1991[1]
Portugal	52.3–64.1	Roque
Portugal	41.7	Fleischer and Fleischer 1992
France	39.9	Baritaux *et al.*, 1992
Portugal	38.2	Baritaux *et al.*, 1992
Bulgaria	59.6	Baritaux *et al.*, 1992
Egypt	41.1	Baritaux *et al.*, 1992
The Netherlands	not mentioned	Suchorska and Osinska 1992
Egypt	51.6	Staath and Azzo 1993
Turkey	43.7	Pérez-Alonso *et al.*, 1995
Turkey	17.7–24.3	Özek *et al.*, 1995

[1] enantiomeric distribution of linalool: R-(–)(90.7–100%), S-(–) (0–9.3%).

O. canum originating from Aruba contained 7.0% of 1,8-cineole (Fun and Baerheim Svendsen 1990).

Charles *et al.* (1990) found 1,8-cineole to be the main compound together with eugenol in an essential oil of a Peruvian *O. micranthum*. The 1,8-cineole content was highest in the leaf oil (20%). The flower oil contained 7.0% and the stem oil 10.9% 1,8-cineole. An Egyptian *O. rubrum* oil contained rather high amounts of 1,8-cineole. In the leaf oil the 1,8-cineole concentration was 16.5% and in the flower oil it was 14.5% (Karawya *et al.*, 1974).

The content of 1,8-cineole in the essential oil of *O. basilicum* varies from 2 to 16%, being generally less than 10%. The oil of *O. basilicum* grown in Madagascar (Randrimiharisoa *et al.*, 1986), Nigeria (Ekundayo *et al.*, 1987), Italy (Tateo 1989), Taiwan (*O. basilicum* var. *minimum*) (Cheng and Liu 1983) and in Turkey (Akgul 1981) contained less than 5% of 1,8-cineole. In addition, *O. basilicum* oils originating from Haiti (Srinivas 1986), Taiwan (Sheen *et al.*, 1991), Israel (Fleisher 1981, Fleisher and Fleisher 1992), Turkey (Perez-Alonso 1995), Egypt (Karawya *et al.*, 1974) contained less than 10% 1,8-cineole.

Among the major constituents of basil oils the content of (*Z*)-β-ocimene + 1,8-cineole varied from traces to 13.6% (Lawrence 1980). Özek *et al.* (1995) compared the chemical composition of *O. basilicum* essential oils of Turkish origin obtained either by steam distillation or by hydrodistillation. They found that the oil obtained by steam distillation contained only about half the amount of 1,8-cineole compared to the oil obtained by hydrodistillation. The content of 1,8-cineole was 7.0% and 13.6%, respectively. Also Laakso *et al.* (1990) found the same phenomenon when studying the difference in chemical composition of essential oils obtained either by steam distillation or by hydrodistillation. They reported that the steam-distilled oil from *O. sanctum* of German origin contained 5.6% 1,8-cineole and the oil obtained by hydrodistillation from the same material contained 11.0% 1,8-cineole.

Modawi *et al.* (1983) reported 1,8-cineole to be one of the main compounds in the essential oil of *O. basilicum* var. *thyrsiflorum* grown in Sudan and Holm *et al.* (1989) reported a new chemotype of *O. basilicum*. They found 1,8-cineole to be the main compound in the material originating from Hungary and cultivated in Finland.

Citral, citronellal and geraniol

Citral, citronellal and geraniol are acyclic monoterpenes which generally co-exist in plant volatiles. Citral is a mixture of two acyclic monoterpene aldehydes, geranial [(*E*)-citral] and neral [(*Z*)-citral]. According to Hegnauer (1966) the citral-type essential oil of *O. canum* contains about 75% citral, *O. gratissimum* about 65%, *O. menthaefolium* Hochts. var. *citrata* about 56%, *O. canum* about 70% and the oil of *O. pilosum* Roxb. 34% of citral. Co-occurence of geraniol with citral has been detected in oils of *O. gratissimum* (geraniol 25%) and *O. menthaefolium* var. *citrata* (geraniol 10%). Hegnauer (1966) and Hoppe (1975) reported a high content of citronellal (41%) together with citral in an oil of *O. pilosum* .

Five chemotypes of *O. basilicum* have been reported. One of these was a geraniol rich chemotype (40–50%) and another chemotype contained 20–30% geraniol together with linalool and eugenol (Sobti and Pushpangadan 1982).

Thymol

The isolation of thymol from *O. basilicum* was reported for the first time in 1988 by Fatope and Takeda. Thymol is more characteristic of the essential oil of *O. gratissimum* where it is found in amounts of 19.3–47.6% (Hegnauer 1966, Sofowora 1970, Sainsbury and Sofowora 1971, Ntezurubanza *et al.*, 1987, Dro 1974, Pino *et al.*, 1996) and *O. viride* (40–65%) (Hegnauer 1966, Hoppe 1975, Simon *et al.*, 1990).

Sesquiterpenes

The sesquiterpenes in the essential oils of *Ocimum* species have not been studied as intensively as the monoterpenes and the phenylpropanes. However, quite many sesquiterpene hydrocarbons and oxygenated sesquiterpenes have been identified in the *Ocimum* oils (Table 4.1).

Lawrence (1971) identified traces of sesquiterpene hydrocarbons from an essential oil of *O. basilicum* and later one more sesquiterpene hydrocarbon and two oxygenated sesquiterpenes (Lawrence 1978). The identified compounds were α-amorphene, bicycloelemene, bicyclogermagrene, *cis*-α-bergamotene, *trans*-α-bergamotene, ε-bulgarene, β-bourbonene, α-cadinene, γ-cadinene, δ-cadinene, T-cadinol, calamene, caryophyllene oxide, β-cedrene, 2-*epi*-α-cedrene, β-cubebene, β-elemene, δ-elemene, α-farnesene, germacrene-D, α-guaiene, δ-guaiene, γ-gurjunene, α-humulene, isocaryophyllene, α-muurolene, α-selinene and β-selinene. In addition to these sesquiterpenes, Nykänen (1987) identified β-bisabolene, α-bulnesene, β-cadinene, β-caryophyllene, β-elemene, β-farnesene, germacrene-B and γ-muurolene from an essential oil of *O. basilicum* grown in Finland.

Karawya *et al.* (1974) identified two oxygenated sesquiterpenes (nerolidol and farnesol) from an essential oil of *O. basilicum* and *O. rubrum*. In the same year Terhune *et al.* (1974) reported 1-epibicyclosesquiphellandrene as a trace compound in basil oil. Sainsbury and Sofowora (1971) identified sesquiterpenes, including α-cubebene, copaene, β-elemene, caryophyllene, *allo*-aromadendrene, α-humulene, β-selinene and δ- and γ-cadinene, from a Nigerian *O. gratissimum* oil. Later on Kekelidze and Beradze (1975) found that an eugenol-type basil oil, presumed to be from *O. gratissimum*, also contained sesquiterpene hydrocarbons (α- and β-caryophyllene and α- and β-santalene). Since that lists of identified sesquiterpenes have been given in numerous publications (*cf.* Lawrence 1978, 1980, 1982, 1986a, 1986b, 1987, 1988, 1989, 1992a, 1992b, 1995, 1997, Carmo *et al.*, 1990, Sheen *et al.*, 1991, Simon *et al.*, 1992, Fleisher and Fleisher 1992, Özek *et al.*, 1995, Perez-Alonso *et al.*, 1995).

Gaydou *et al.* (1989) developed a fractionation method for basil oil (*O. basilicum*) obtained from plants grown in Madagascar for the identification of sesquiterpenes. They succeeded in identifying altogether 31 sesquiterpene hydrocarbons from this oil, eleven of those for the first time from basil oil. New compounds identified were α-gurjunene, scapanene, α-elemene, *allo*-aromadendrene, (*E*)-α-bisabolene, viridiflorene, bicyclogermacrene, (*Z*)-α-bisabolene, γ^2-cadinene, α-cadinene and α-calacorene. β-Caryophyllene (34.6% of the sesquiterpene fraction) and β-elemene (18.0%) were identified as the main sesquiterpenes.

Hasegawa *et al.* (1997) examined essential oils produced from 9 different cultivars of *O. basilicum* by gas chromatography – mass spectrometry (GC-MS) and identified

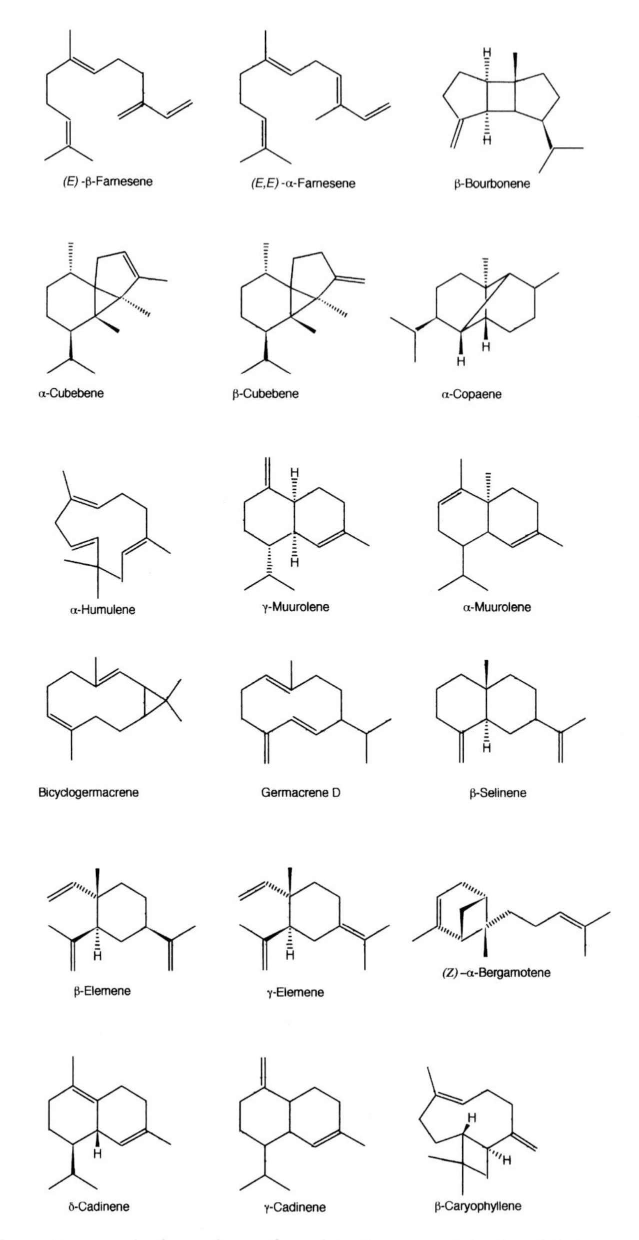

Figure 4.2 Sesquiterpene hydrocarbons found in the essential oils of *Ocimum* species.

in all 19 sesquiterpene hydrocarbons and 9 oxygenated sesquiterpenes. Of the sesquiterpenes β-caryophyllene had the highest amount, varying from 4.0 to 9.1% of the total oil. The content of the other sesquiterpenes together was less than 5%. In general, sesquiterpenes are trace or minor constituents in *O. basilicum* essential oil.

An essential oil of *O. micranthum* originating from Peru has been found to be exceptionally rich in sesquiterpenes. The total sesquiterpenes accounted for 48.4, 85.8 and

78.5% of the oil in leaves, flowers and stems, respectively. The major sesquiterpenes were β-caryophyllene (18.9–19.3% of the total oil), β-selinene (4.7–14.0 %) and a γ-elemene isomer (14.4–28.2%). Other identified sesquiterpenes were two other γ-elemene isomers, γ-copaene and two β-elemene isomers (Charles *et al.*, 1990).

O. canum essential oil contains both sesquiterpene hydrocarbons and oxygenated sesquiterpenes, together 15%, including δ-cadinene (traces–9.8 %) (Ekundayo *et al.*, 1989, Ntezurubanza *et al.*, 1985, Fun and Baerheim Svendsen 1990), bergamotene (2.6–8.2%) (Ntezurubanza *et al.*, 1985, Fun and Baerheim Svendsen 1990), β-caryophyllene (2.0–5.3%) (Xaasan *et al.*, 1981, Ntezurubanza *et al.*, 1985, Ekundayo *et al.*, 1989), smaller amounts of α- and β-selinene (1.5–3.0%) (Xaasan *et al.*, 1981, Ekundayo *et al.*, 1989), γ-cadinene (1.9%) (Fun and Baerheim Svendsen 1990) and traces of α-cubebene, α-copaene, β-bourbonene, β-cubebene, β-elemene (4.3%) (Ekundayo *et al.*, 1989), α-humulene, germacrene-D (2.1%) (Ekundayo *et al.*, 1989), β-caryophyllene oxide, bergamotene, (*E*)-nerolidol, cubenol, T-cadinol, α-cadinol and α-bisabolol (Xaasan *et al.*, 1981, Ntezurubanza *et al.*, 1985, Ekundayo *et al.*, 1989, Fun and Baerheim Svendsen 1990).

Sesquiterpenes have been found in essential oils of *O. gratissimum* grown in China (Yu and Cheng 1986, Wu *et al.*, 1990), in Russia (Zamureenko *et al.*, 1986), in India (Khanna *et al.*, 1988), in Brazil (Maia *et al.*, 1988, Vostrowsky 1990), in Aruba (Fun and Baerheim Svendsen 1990), in Madagascar (De Medici *et al.*, 1992), and in Cuba (Pino *et al.*, 1996).

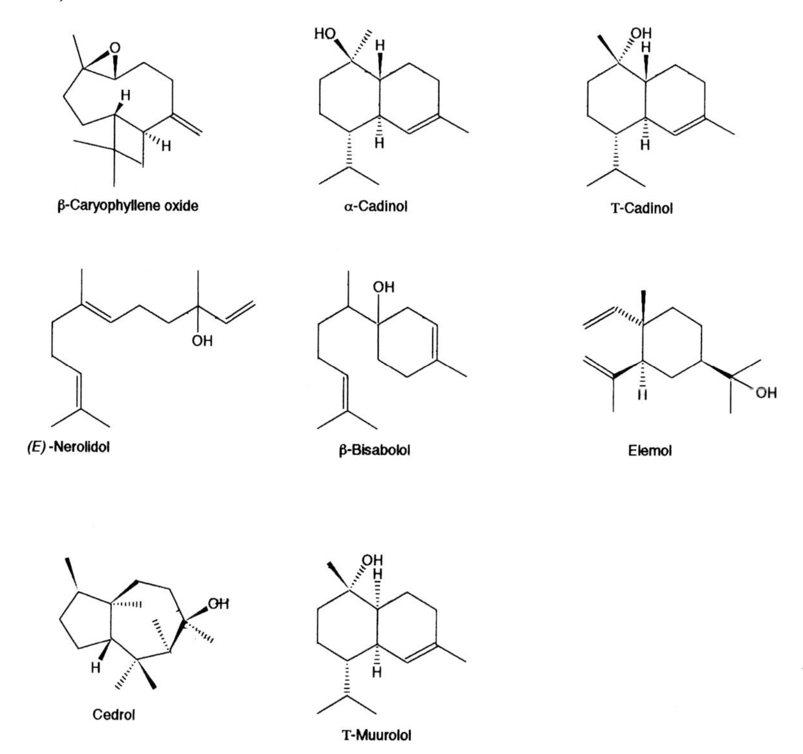

Figure 4.3 Oxygenated sesquiterpenes found in the essential oils of *Ocimum* species.

Fun and Baerheim Svendsen (1990) analyzed three *O. gratissimum* samples from Aruba. The essential oil of one sample contained in addition to a phenylpropane derivative (eugenol) also 45.5% β-caryophyllene and 11.8% β-elemene. The two other samples contained 36.0–39.8% β-caryophyllene and 9.0–10.4% β-elemene. High contents of β-caryophyllene in *O. gratissimum* oils are reported by Pino *et al.* (1996) and Lawrence (1992), 7.2–11.5% and 2.5–27.3%, respectively. Zamureenko *et al.* (1986) reported that the amount of germacrene-D can vary from 1.8% to 15.7% in a Russian *O. gratissimum* oil and β-caryophyllene from 1.2% to 6.5%. Maia *et al.* (1988) found a high content of (*E*)-β-farnesene (19.0%) in a Brazilian *O. gratissimum* oil. Lawrence (1992) reported 9.2% bicyclogermacrene in a chemotype of *O. gratissimum.*

The sesquiterpenes reported in the essential oil of *O. gratissimum* are *allo*-aromadendrene, bicyclogermacrene, β-bisabolene, (*Z*)-α-bisabolene, α-bisabolol, *trans*-α-bergamotene, β-bourbonene, cadinene, α-cadinene, γ-cadinene , δ-cadinene, α-cadinol, cadinenol, T-cadinol, β-caryophyllene, caryophyllene oxide, cedrol, α-copaene, α-cubebene, β-cubebene, β-elemene, δ-elemene, elemol, α-farnesene, (*E*)-β-farnesene, germacrene-D, α–humulene, isocaryophyllene, α-muurolene, γ-muurolene, T-muurolol, (*E*)-nerolidol and β-selinene.

The essential oil of *O. trichodon* growing wild in Rwanda contains eugenol as a main compound (44–81%) and small amounts of sesquiterpenes. From the leaf and flower oil α-copaene, β-elemene, β-caryophyllene (2.0–8.6%), α-bergamotene isomer, α-humulene, germacrene-D (1.2–10.0%), α-farnesene (1.6–5.2%), δ-cadinene, γ-cadinene and β-caryophyllene oxide have been identified (Ntezurubanza *et al.*, 1985).

The leaf oil essential oil of *O. keniense* contains high amounts of 1,8-cineole (about 38%) and methyl chavicol (about 24%). In addition some sesquiterpenes, such as *trans*-caryophyllene, β-elemene, *allo*-aromadendrene (4.6%), calarene, β-maaliene (3.4%) and traces of elemol and cedrol, have been found (Mwangi *et al.*, 1994).

Very high contents of sesquiterpenes in an essential oil of *O. sanctum* grown in Germany have been reported. In a steam distilled oil 19.6% α-bisabolene and 15.4% β-bisabolene were detected. In addition to that also (*E*)-β-bergamotene, β-caryophyllene, (*E*)-β-farnesene, α-humulene and α-bisabolol have been found (Laakso *et al.*, 1990).

Phenylpropane Derivatives as Main Constituents

Biosynthetically phenylpropanes are formed in plant cells via the shikimic acid pathway. L-Phenylalanine and L-tyrosine which are formed this way serve as precursors for a wide variety of natural products, including the phenylpropanes in the essential oils. Phenylpropanes are synthesized from *trans*-cinnamic acid [(*E*)-cinnamic acid] which in turn is generated by the elimination of ammonia from the side-chain of L-phenylalanine. Many characteristic compounds of basil oil, including methyl cinnamate, methyl chavicol (estragol), eugenol and methyl eugenol originate from (*E*)-cinnamic acid.

Methyl chavicol

The European type of basil (*Ocimum basilicum*) oil as well as the Egyptian and Reunion or Comoro basil oils are characterized by high concentrations of methyl chavicol,

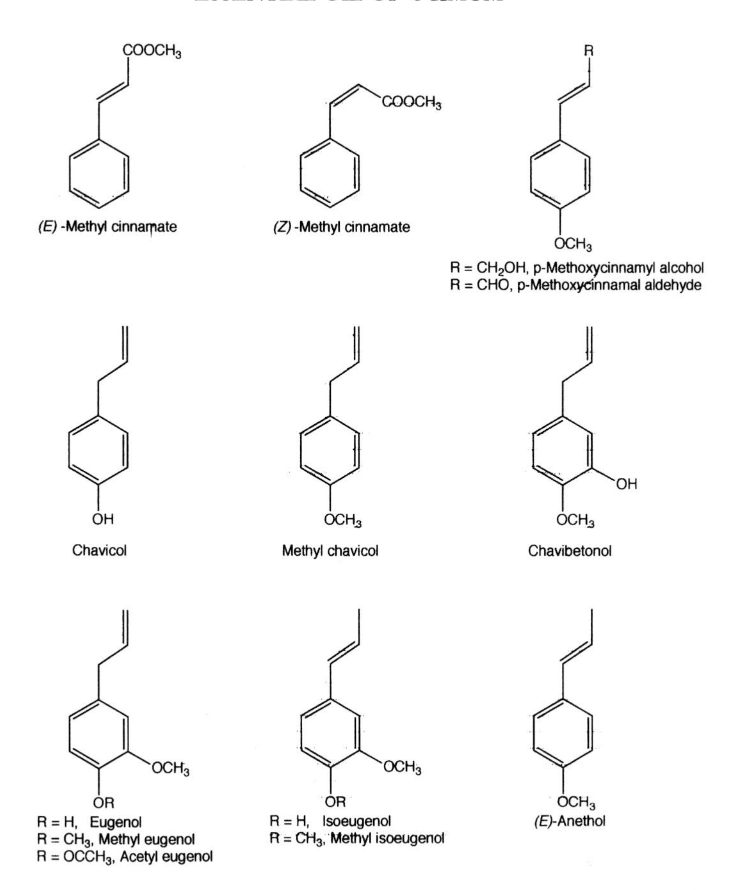

Figure 4.4 Phenylpropane derivatives found in the essential oils of *Ocimum* species.

also known as estragol. An Egyptian basil oil contained 81.1% of methyl chavicol (Staath and Azzo 1993).

According to Günther (1949) and Lawrence (1978, 1989) the basil oils distilled in France, Italy, Bulgaria, Egypt, Hungary, South Africa and occasionally also in the United States contain approximately equal proportions of methyl chavicol and linalool. In contrast Reunion or Comoro basil oil contains little if any linalool, but has a very high concentration of methyl chavicol (Lawrence 1980, Simon *et al.*, 1990). Boniface *et al.* (1987) found that a commercial oil of Asian or Oriental origin, including the oils from Reunion and Comoro, contained 89.0% methyl chavicol. Earlier Conan (1977) compared the basil oils from Reunion, France and Morocco and found that the oil from Comoro consisted mostly of methyl chavicol (85.8%). Also basil grown in Madagascar and Pakistan has been shown to be rich in methyl chavicol (Lawrence 1992)(Table 4.4).

Methyl cinnamate

Methyl cinnamate is the main constituent of *O. basilicum* and *O. canum* oils. *O. basilicum* originating from various tropical countries of the world (*e.g.* Guatemala, Haiti, India and some African countries) has methyl cinnamate as a main constituent in their essential oils (Lawrence 1978). The amount of methyl cinnamate ranged from traces to 15.5% in a morphologically heterogenous basil material (Lawrence 1980). Later on in 1986 Brophy and Jogia examined the chemical composition of two different chemotypes of *O. basilicum* grown in Fiji. One of the chemotypes had methyl eugenol as a main compound (24.7%) but was also rich in (*E*)-methyl cinnamate (23.6%). The other chemotype was very rich in (*E*)-methyl cinnamate (64.5–68.6%) (Table 4.5).

Hasegawa *et al.* (1997) detected a methyl cinnamate level of 73.7 % in an essential oil of *O. basilicum* collected from the Philippines. They did not separate the (*Z*)- and

Table 4.4 The methyl chavicol content of basil oils

Origin	*Methyl Chavicol Content (%)*	*References*
Calabria	29.8–35.9	Günther 1949
"exotic"	87.8	Lawrence *et al.*, 1971
Thailand	88.2	Lawrence *et al.*, 1972
Italy, France, Morocco	32	Lawrence 1973
	55–85	Hoppe 1975
	68.2–86.7	Gulati 1977
South Africa, Egypt, India	2.4–27.7	Peter and Rémy 1978
Yugoslavia, France		
Pakistan, the Comoros	65.7–85.8	Peter and Rémy 1978
Israel	1.5–28.9	Fleischer 1981
	0.3–88.6	Lawrence *et al.*, 1981
Sudan	40–70	Modawi *et al.*, 1983
Taiwan	35.0 (leaf)	Cheng and Liu 1983
	41.8 (flowers)	Cheng and Liu 1983
	4.9 (stem)	Cheng and Liu 1983
the Comoros, Pakistan	74.3–85.5	Vernin *et al.*, 1984
Vietnam, Madagascar	81.6–86.2	Randriamiharisoa and Gaydou 1985
Cuba	56.6	Tapanes *et al.*, 1985
Haiti	26.8	Srinivas 1986
Madagascar	73.9–87.3	Randriamiharisoa *et al.*, 1986
Finland	n.d.–77.1	Nykänen 1987
see below[1]	74.0–89.0[2]	Boniface *et al.*, 1987
	68.9–88.6[3]	Boniface *et al.*, 1987
Nigeria	84.1	Ekundayo *et al.*,[4] 1987
U.S.A.	5–47	Simon *et al.*, 1990

[1] Boniface *et al.* (1987): Origin of the basil oils; Group 1(Asian or Oriental). Thailand, Reunion, Comoro Islands, India, Pakistan, Madagascar, France; Group 2 (Europe). Italy, Yugoslavia; Group 3 (African). Morocco, South Africa, Egypt.

[2] commercial oil

[3] experimental oil

[4] Ekundayo *et al.* (1987) identified for the first time 5-methyl-2-heptanone, neryl acetate, β-sesquiphellandrene and cubenol in basil oil.

Table 4.4 (continued)

	83.1	Chien 1988
Comorro	75.1–87.4	Lawrence 1988
Finland	17–77	Nykänen 1989
Madagascar	74–87	Gaydou 1989
Taiwan, Taichung	68.5–86.6	Sheen *et al.*, 1991
U.S.A., Indiana	30.3	Simon *et al.*, 1992
France	2.8–86.3	Baritaux *et al.*, 1992
	not mentioned	Suchorska and Osinska 1992
Egypt	81.1	Staath and Azzo 1993
France, West-Africa, India (Kerala)	55[6]	Gupta 1994
	60–75[7]	Gupta 1994
Bhutan	89.3	Lawrence 1995
the Philippines, U.S.A., CA	tr.–82.8	Hasegawa *et al.*, 1997

[5] Baritaux *et al.* (1992): Geographical origin of the seeds was France, Portugal, Bulgaria, Finland, Egypt, Nigeria (Ife), the Comoros, Madagascar, Cuba (Havana) and India. Four chemotypes were detected as follows: chemotype A (linalool), chemotype B (methyl chavicol), chemotype AB (linalool>methyl chavicol) and chemotype BA (methyl chavicol>linalool).

[6] original plant

[7] self-pollinated generation

Table 4.5 Relative amounts of (*Z*)- and (*E*)-methyl cinnamate in basil oils

Origin	*(Z)–Methyl Cinnamate (%)*	*(E)–Methyl Cinnamate (%)*	*References*
Turkey	4.6	27.3	Perez-Alonso *et al.*, 1995
Fiji	4.5	23.6	Brophy and Jogia 1986
Fiji	7.6–11.2	64.5–68.6	Brophy and Jogia 1986
Haiti	5.7	22.5	Srinivas 1986
Morocco	nd.[1]	38.5	Lawrence 1989
Finland	nd.–1.7	nd.–12.0	Nykänen and Nykänen, 1987
Israel	4.7	31.9	Fleisher and Fleisher, 1992
Egypt	0.8	4.8	Staath and Azzo 1993
Turkey	1.5	12.7	Özek *et al.*, 1995
Portugal		tr.[1]–60.4	Lawrence 1988
	Methyl cinnamate (%)		
Haiti	27.0		Günther 1949
Calabria	15.0		Günther 1949
	15.0–75.0		Hegnauer 1966
Comorro	54.2–56.3		Gildemeister and Hoffmann 1961
India	57.0–70.0		Hoppe 1975
India	60.0–65.0		Sobti and Pushpangadan 1982

[1] nd = not detected, tr. = traces

(*E*)-forms of methyl cinnamate. Also Lawrence (1988) reported a high concentration of (*E*)-methyl cinnamate (64.0%) in one of the *O. basilicum* selections (Tables 4.6–4.7).

Eugenol

Eugenol has been determined to be a predominant constituent in certain essential oils of *O. basilicum*. Günther (1949) differentiated basil oils into four chemotypes and one of those was an eugenol-type (Table 4.6). *O. basilicum* grown in Nigeria contains eugenol as one of the major constituents (16.9%) (Ekundayo *et al.*, 1987). *O. basilicum* grown in Taiwan contains 53.1%(Cheng and Liu 1983), in Portugal 9.7–17.6% (Roque 1991), and in Israel 4.0–27.9% eugenol (Fleisher 1981).

Eugenol is very common as a constituent of *O. gratissimum* essential oil. It has been detected as a main compound in oils from Russia, North America, China, India and Madagascar (Tables 4.12 and 4.13). The eugenol content usually ranges from 40 to 90% when it is a main constituent. The strain Clocimum, obtained as a result of a screening and breeding programme of *Ocimum* species in India, can contain 60–70 % of eugenol (Sobti *et al.*, 1982)

Janssen *et al.* (1989) examined the essential oils of *O. canum*, *O. gratissimum*, *O. trichodon* and *O. urticifolium* (syn. *O. suave*) grown in Rwanda and found that the essential oil of *O. gratissimum* contained 10.7% eugenol, but the oils of *O. trichodon* and *O. urticifolium* as much as 74.7% and 37.6–71.0%, respectively. Also Chogo and Crank (1981) found a high amount of eugenol (71.5%) in the essential oil of *O. suave* grown in Tanzania.

Ekundayo *et al.* (1989) studied the essential oil from leaves of *O. canum* collected in the Ibadan area (Nigeria) and detected eugenol as the main constituent (66.4%). The essential oil of leaves and flowers of *O. trichodon* Baker ex Guerke grown in Rwanda is rich in eugenol (44.2–81.2%) (Ntezurubanza *et al.*, 1986).

The essential oil of *O. sanctum* can contain eugenol about 70% (Hegnauer 1966). The leaf essential oil of *O. micranthum* contains 20.5% of eugenol, but only traces of eugenol can be found in the oils obtained from the flowers and stems (Charles *et al.*, 1990). (Tables 4.6–4.8, 4.10 and 4.12–4.14).

Methyl eugenol

Methyl eugenol is not very common in higher concentrations in the essential oils of *Ocimum* species. However, 20% methyl eugenol has been found in the oil of *O. sanctum* (Hegnauer 1966). Methyl eugenol has been detected as the main compound

Table 4.6 Chemotypes of sweet basil oil (Guenther 1949)

Type of Oil	*Principal Constituent(s)*
I European type	methyl chavicol, linalool (no camphor)
II Réunion type	methyl chavicol, camphor (no linalool)
III Methyl cinnamate	methyl chavicol, linalool, methyl cinnamate
IV Eugenol type	eugenol

Table 4.7 Five chemotypes of basil according to Sobti and Pushpangadan (1982)

Chemotype	*Range of Major Compounds (%)*					
1	geraniol	40 – 50	eugenol	20–30		
2	eugenol	20 – 40				
3	camphor	10 – 15				
4	methyl cinnamate	60 – 65				
5	geraniol	20 – 30	linalool	30–35	eugenol	20–30

in an essential oil of *O. basilicum* var. *minimum* of Taiwanese origin (42.9–64.3%) and in the stem oil of *O. basilicum* (53.1%) of the same origin (Cheng and Liu 1983). In an oil of *O. gratissimum* of Brazilian origin, 46.8% methyl eugenol was found (Vostrowsky 1990). A high methyl eugenol concentration (24.7%) was reported by Brophy and Jogia (1986) from plant material of *O. basilicum* grown locally in Fiji.

Methyl eugenol (traces to a few percent) has been found in the essential oils of the following *Ocimum* species: *O. canum* (Ekundayo *et al.*, 1989), *O. basilicum* var. *canum* (Fun and Baerheim Svendsen 1990), *O. gratissimum* (Sainsbury and Sofowora 1971, Khanna *et al.*, 1988, Ntezurubanza *et al.*, 1989 (*trans*-methylisoeugenol), *O. basilicum* (Karawya *et al.*, 1974, isoeugenol; Masade 1976, Randriamiharisoa and Gaydou 1985, Randriamiharisoa *et al.*, 1986, Ekundayo *et al.*, 1987; Nykänen and Nykänen 1987, Tsai and Sheen 1987, Chien 1988, Lawrence 1988, Chien 1988, Akgul 1981, Tateo *et al.*, 1989, Kostrzewa and Karwowska 1991, Mariani *et al.*, 1991, Sheen *et al.*, 1991, Bobin *et al.*, 1991, Perez-Alonso *et al.*, 1995), *O. rubrum* (Karawya *et al.*, 1974, isoeugenol) *O. sanctum* (Lai *et al.*, 1978) and *O. gratissimum* (Ntezurubanza *et al.*, 1987, Khanna *et al.*, 1988, Fun and Baerheim Svendsen 1990, Lawrence 1992).

CHEMOTYPES OF *OCIMUM BASILICUM*

In many studies it is pointed out that *O. basilicum* occurs in numerous subspecies, varieties and forms, which differ from each other both in the morphology and in the chemical composition of the essential oil. As early as 1930 Guillaumin classified basil (*O. basilicum*) oil into four categories:

(a) Common basil oil has linalool and methyl chavicol as main compounds, cineole and eugenol also exist but camphor and methyl cinnamate are absent
(b) Camphor-type basil oil is rich in camphor but it also contains smaller amounts of α-pinene, cineole, linalool and methyl chavicol
(c) Methyl cinnamate-type basil oil consists of 15–75% methyl cinnamate
(d) Eugenol-type basil oil is rich in eugenol (30–80%).

According to Günther (1949) it is possible to differentiate basil oils into four types (Table 4.6).

The European basil oil is commonly divided into two subtypes, a French sweet basil oil and an American sweet basil oil.

Sobti and Pushpangadan (1982) reported the existence of five different chemotypes of basil, including a new eugenol chemotype (Table 4.7). On the basis of more than 200 analyses of basil oils, Lawrence (1988, 1992) divided *O. basilicum* into four major essential oil chemotypes[1], each with a number of small variants: (1) methyl chavicol-rich, (2) linalool-rich, (3) methyl eugenol-rich and (4) methyl cinnamate-rich (Table 4.8). The analysed oils were categorized into four chemotypes (chemovars) for either mevalonic acid pathway or shikimic acid pathway alone or four chemotypes with dual biosynthetic pathways. Lawrence (1988) did not find any eugenol-rich. According to Lawrence the strain reported by Sobti and Pushpangadan (1982), which can be thought of as a fifth chemotype, must be considered to be somewhat speculative. Because eugenol contents higher than 14 % have already been encountered, Lawrence thought that the likelihood of an eugenol-rich chemotype existing seemed quite probable. Later on Pushpangadan and Bradu (1995) classified basil chemotypes in a similar way as Lawrence but included an eugenol-rich oil in its own chemotype of *O. basilicum* like Sobti and Pushpangadan (1982).

Boniface *et al.* (1987) applied discriminant analysis to differentiate between basil oils of different origin. They found that oils from Thailand, Reunion, Comoro Islands, India, Pakistan, Yugoslavia, Morocco, South Africa and Egypt could be classified by the levels of 1,8-cineole + *cis*-β-ocimene, linalool, methyl chavicol, methyl cinnamate and eugenol into three categories: Group 1. Asian (Oriental) oils, group 2. European oils and group 3. African oils. The Asian oils are characterized by high concentrations of methyl chavicol (68.9–89.0%) and by smaller amounts of linalool (0.5–16.7%). The European oils are rich in linalool (23.0–75.4%) and methyl chavicol (0.4–43.6%) and also have rather high concentrations of methyl cinnamate (0.1–15.5%) and 1,8-cineole + *cis*-β-ocimene (2.7–13.6%). The African oils differ from the Asian and European oils by having a high level of eugenol (5.9–19.2) and lower levels of methyl chavicol (2.4–26.6%).

Baritaux *et al.* (1992) summarized the reported data of basil essential oils from France, Portugal, Bulgaria, Finland, Egypt, Nigeria, Comoro Islands, Madagascar, Cuba and India. They formulated four major chemotypes according to the essential oil composition and major compounds (Table 4.9).

A 1,8-cineole rich (about 30%) basil oil, indicating a new chemotype of basil, was found when 14 basil (*O. basilicum*) cultivars of different origins grown in Finland were examined. In this study the volatile oil analyses were carried out by headspace gas chromatography (HSGC). The identification was made by analyzing the hydrodistilled oil by gas chromatography – mass spectrometry (GC/MS). Altogether 49 components were identified. On the basis of the main compounds the cultivars were classified as follows: (1) linalool-methyl chavicol-type (4 cultivars), (2) methyl cinnamate-type (3 cultivars), (3) linalool-type (4 cultivars), (4) linalool-eugenol-type (1 cultivar) and (5) 1,8-cineole-type (2 cultivars) (Holm *et al.*, 1989).

---

[1] The chemical diversity of basil oils, derived from different chemotypes and cultivars, is thought to be due to polymorphism in basil, which is caused by interspecific hybridization. Apart from this the species can be classified into four distinct chemotypes, with many subtypes based on biosynthetic pathways of major compounds in the oils (Lawrence 1988).

Table 4.8 Chemotypes of *Ocimum basilicum* with dual biosynthetic pathways (Lawrence 1988)

Type of Oil	*Principal Constituent(s)*
Type 1.1.	methyl chavicol > linalool
1.2.	methyl chavicol > linalool > methyl cinnamate
Type 2.1.	linalool > methyl chavicol
2.2.	linalool > methyl chavicol > eugenol
2.3.	linalool > eugenol > methyl eugenol
2.4.	linalool > methyl cinnamate
Type 3.1.	methyl eugenol > linalool > eugenol
Type 4.1.	methyl cinnamate > methyl chavicol

Ruberto *et al.* (1991) identified dihydrotagetone as the main component (82%) in the steam distilled volatile oil of *Ocimum basilicum* var. *hispidum* (Lam.) Chiov. In this steam-distilled oil they identified a total of 21 compounds, representing about 90% of total oil. The oil contained only traces of linalool. Phenylpropane derivatives were not found in this oil.

Lawrence (1992) has criticized chemotypic classification in cases when cultivars contain at least two major compounds. The reason for this criticism is the fact that the essential oil composition of basil oils can be different before and after drying as has been shown in many studies. Baritaux *et al.* (1992) studied the effects of drying and storage on the essential oil of basil. They found that the content of methyl chavicol and eugenol decreased drastically after drying (45°C for 12 hours) and storing (3, 6 and 7 months) while the content of linalool and 1,8-cineole increased over the same time period. The essential oil yield decreased from 312.8 mg to 108.2 mg per 100 g of dry weight. The loss in total oil yield after 7 months storage was 66%. The percentage proportions of 1,8-cineole and linalool increased about two times (from 4 to 9% and from 24 to 45%, respectively). The percentage of methyl chavicol decreased from 55% to about 11% and eugenol from 7% to 3%. This means that the fresh plants represented a clear cut methyl chavicol chemotype while the dried and stored plants represented a linalool type. Not only the contents of 1,8-cineole and linalool but also the contents of sesquiterpene hydrocarbons such as *trans*-β-bergamotene, γ-muurolene and cadinene isomers increased on drying. Baritaux *et al.* (1992) concluded that the essential oil loss was due mainly to evaporation and the increase of oxygenated monoterpenes was due to hydrolysis of monoterpene glucosides during distillation. Nykänen and Nykänen (1987) studied the variability

Table 4.9 Four chemotypes of *O.basilicum* according to Baritaux *et al.* (1992)

Chemotype	*Major Compound(s) (%)*	*Seed Origin*
A	linalool (59.6)	Bulgaria
B	methyl chavicol (84.1–86.3)	Nigeria, Comorro Islands
AB	linalool (39.9–41.1) > methyl chavicol (23.4–28.2)	France, Egypt
BA	methyl chavicol (56.6–77.5) > linalool (10.0–19.0)	India, Cuba, Finland

and differences in the chemical composition of basil oil before and after drying the material. They reported about a 40% loss of volatile oil on drying. They also detected a significant decrease in the levels of methyl chavicol and eugenol, while the levels of 1,8-cineole and linalool tended to increase.

Differences in the chemical composition of basil oil in fresh and dried basil were also shown by Grayer *et al.* (1996). They reported that the essential oil obtained from fresh leaves of an *O. basilicum* cultivar had methyl chavicol as the main compound (53.4%), but the essential oil of the same cultivar distilled from the dried leaves contained linalool as the main compound (46.7%) and the proportion of methyl chavicol was only 29.0%. Grayer *et al.* (1996) also showed that the fresh material of two other cultivars (147/97 and 76) could have been considered to belong to the methyl chavicol-rich and eugenol-rich groups, respectively, but the dried leaves of the same cultivars would have had to be classified as methyl eugenol-rich and linalool-rich, respectively. For that reason Grayer *et al.* proposed that chemotypes should be based not only on one major compound because frequently there are two or more major compounds which may be present in almost equal amounts. It would be better to determine chemotypes by assigning an essential oil profile that is based on all the major compounds *e.g.* compounds constituting more than 20% of the total oil. Apart from using fresh or dried plant material there are many other factors affecting the proportional essential oil composition. Some factors, such as the extraction method used, the age (phenological stage), the organ of the plant used for study, environmental conditions during the vegetation period, use of fertilizers and covering the plants, are also important determinants of the essential oil composition.

CHEMOTYPES OF *OCIMUM CANUM*

The essential oil of *O. canum*, also known as camphor basil, is rich either in oxygenated monoterpenes (camphor, citral, linalool) or in phenylallyl ethers such as eugenol, methyl chavicol and methyl cinnamate (Table 4.10). Hegnauer (1966) reported three chemotypes of *O. canum* Sims.: (a) camphor-type (about 65% of d-camphor), (b) methyl cinnamate-type (about 55%) and (c) citral-type (about 75%). Camphor- and methyl cinnamate-type oils are also rich in linalool (25%), citral-type oils also contain limonene and geraniol (35%) (Gildemeister and Hoffmann 1961). Both the *E*- and *Z*-forms of methyl cinnamate have been detected in the methyl cinnamate-type oil (Table 4.5).

The essential oil of *O. basilicum* var. *canum* (syn. *O. canum*) grown in Aruba contains *Z*-methyl cinnamate (*cis*-methyl cinnamate) from 38.1 to 64.5% and *E*-methyl cinnamate (*trans*-methyl cinnamate) from 5.3 to 8.2%. 1,8-Cineole (7.0%) was reported to be one of the key compounds in this oil (Fun and Baerheim Svendsen 1990).

A leaf oil of *Ocimum canum* grown in the Eleiyele area in Ibadan (Nigeria) contained eugenol as its main compound (66.4%) and some minor components such as δ-cadinene (9.8%), β-elemene (4.3%), β-caryophyllene (3.5%), α-selinene (2.1%), germacrene-D (2.1%) and β-selinene (1.5%). The content of the other identified

Table 4.10 Chemotypes of *Ocimum canum* Sims

Chemotype (Major Compound)	*References*
Methyl cinnamate	Guenther 1949, Hegnauer 1966
Camphor	Guenther 1949, Hegnauer 1966, Gulati *et al.*, 1977, Sobti *et al.*, 1976, Xaasan *et al.*, 1981, Gupta and Sobti 1990, Upadhyay *et al.*, 1991
Citral	Guenther 1949, Hegnauer 1966, Gupta and Sobti 1990
Linalool	Gulati *et al.*, 1977, Sobti *et al.*, 1976, Ntezurubanza *et al.*, 1985, Janssen *et al.*, 1989; Gupta and Sobti 1993, Gupta and Tawa 1997
Eugenol	Ekundayo *et al.*, 1985
Methyl chavicol	Gupta and Sobti 1990

compounds (altogether 29 identified compounds) was less than 1%. A eugenol chemotype of *O. canum* was reported for the first time (Ekundayo *et al.*, 1989). This new eugenol chemotype brought the number of chemotypes of this species to five. Earlier three chemotypes comprising mainly methyl cinnamate, camphor or citral had been recognized (Günther 1949). Cultivation studies carried out in the 1970's in India showed the existence of a linalool-type (89–90%) and a camphor-type (60–80%) (Sobti *et al.*, 1976, Gulati *et al.*, 1977). Later camphor (60%) was found in essential oils of African origin (Xaasan *et al.*, 1981). Also linalool (60–90%) has been reported to be the main compound of African *O. canum* oils (Ntezurubanza *et al.*, 1985, Janssen *et al.*, 1989).

Two chemotypes of *Ocimum americanum* (syn. *O. canum*), one rich in methyl chavicol and the other rich in citral, have been reported (Gupta and Sobti 1990). By reciprocal crossing and studying of the P-, F-1, F-2 and F-3 generations it was shown that the segregation of methyl chavicol and citral was strictly in accordance with the Mendelian principles. Gupta and Sobti (1991) reported the inheritance of linalool and camphor, found as major constituents in the oil of *O. canum*. In this study the linalool content in some of the segregants and the camphor content in certain other segregants was very high, ranging from 75 to 80%. *O. americanum* L. is morphologically an intermediate between *O. canum* and *O. basilicum* (Pushpangadan and Sobti 1982).

The essential oil of *O. americanum* contained as major constituents the following: α-pinene (8.3%), sabinene (8.0%), limonene (7.8%), camphor (26.7%) and γ-selinene (10.9%). The proportion of monoterpene hydrocarbons and oxygenated monoterpenes were 31.6% and 35.4%, respectively. The proportional amount of sesquiterpene alcohols was 23.4% (Upadhyay *et al.*, 1991)

Linalool has been found in the essential oil of *O. canum* from traces to 90% of the total oil. In the linalool-type oil the linalool content ranges from 60 to 90% (Table 4.11).

Table 4.11 The linalool content of the essential oil of *Ocimum canum* Sims

Origin	*Linalool (%)*	*References*
India	63.2	Gupta and Tawa 1997
Rwanda	82.3	Janssen *et al.*, 1989
Rwanda	61.0–88.1	Ntezurubanza *et al.*, 1985
India	89–90	Sobti *et al.*, 1976
	75–80	Gupta and Sobti 1991

CHEMOTYPES OF *OCIMUM GRATISSIMUM*

The essential oil of *O. gratissimum* is characterized either by large amounts of thymol or eugenol (Table 4.12). According to Günther (1949) and Hegnauer (1966), the thymol type does not contain eugenol, and thymol is not present in the eugenol type.

Hegnauer (1966) divided the oils of *O. gratissimum* into three chemotypes:

(a) a thymol-type, which contains 33–44% thymol but no eugenol
(b) a citral-type, which contains mainly citral (65%) and geraniol (25%)
(c) a eugenol-type, which is rich in eugenol (50–90%), but thymol is absent.

According to Günther (1949) and Hegnauer (1966) the oil of *O. gratissimum* can also contain citral as its main component (67%) (citral type) (Table 4.13).

Pino *et al.* (1996) published the composition of the essential oil from the leaves and flowers of *O. gratissimum*. Both the oils was found to be similar, thymol (leaf oil: 19.4%, flower oil: 27.3%) and *p*-cymene (leaf oil: 14.0%, flower oil: 12.8%) and β-caryophyllene (leaf oil: 7.3%, flower oil: 11.5%) were the major compounds in these oils. De Medici *et al.* (1992) examined the essential oil composition of two samples of *O. gratissimum* grown in Madagascar. In both oils eugenol was the main compound (40.3–46.1%). In one of the oils, the content of 1,8-cineole was rather high (12%) and in another oil the levels of some other monoterpenes were more than 10%, *e.g.* γ-terpinene 18.6% and α-terpineol 10.2%. De Medici *et al.* (1992) found 12% of methyl chavicol.

Khanna *et al.* (1988) reported eugenol as the main compound (77.6%) in the essential oil of *O. gratissimum* grown in India. They called this new strain Clocimum,

Table 4.12 The eugenol content in *O. gratissimum* oils

Origin	*Eugenol (%)*	*References*
Russia	31.5–90.5	Zamureenko *et al.*, 1988
North America	85.1	Lawrence 1987
China	84.5	Wu *et al.*, 1990
China	80.8	Yu and Cheng 1986
India (strain: Clocimum)	77.6	Khanna *et al.*, 1988
	21.1–56.7	Lawrence 1992c
Madagascar	40.3–46.1	De Medici *et al.*, 1992
Aruba	20.0	Fun and Baerheim Svendsen 1990
Rwanda	0.3–10.7	Ntezurubanza *et al.*, 1987

Table 4.13 Chemotypes of *Ocimum gratissimum*

Major Compound (Chemotype)	*References*
Thymol	Guenther 1949, Hegnauer 1966, Pino *et al.*, 1996, Ntezurubanza *et al.*, 1987
Citral	Guenther 1949, Hegnauer 1966, Yu and Cheng 1986, Zamureenko *et al.*, 1986, Khanna *et al.*, 1988, Thomas 1989, Wu *et al.*, 1990, Lawrence 1992c, De Medici *et al.*, 1992
Methyl cinnamate	Ntezurubanza *et al.*, 1987, Fun and Baerheim Svendsen 1990
Linalool	Pino *et al.*, 1996
Sesquiterpene hydrocarbon	
-eugenol	Fun and Baerheim Svendsen 1990
p-Cymene	Maia *et al.*, 1988

because the volatile oil obtained had a clove like smell. Ntezurubanza *et al.* (1987) studied an essential oil of *O. gratissimum* growing wild in Rwanda and summarized the literature data on this species. They stated that four chemotypes have been reported: a thymol type containing 33–44% thymol or more than 50% of it; an eugenol type consisting of 62% or 50–90% of this compound; a citral type, containing 67% citral and a methyl cinnamate type, containing 67% of that compound. In their study the oil of *O. gratissimum* cultivated in Rwanda from seeds collected in Cameroon, consisted of 35.4–46.7% thymol and only 0.3% eugenol but the percentage of γ-terpinene was as high as 23%. Fun and Baerheim Svendsen (1990) examined an essential oil of *O. gratissimum* grown in Aruba and found that together with β-caryophyllene, eugenol is one of its major compounds (20.0%).

Sainsbury and Sofowora (1971) reported eugenol to be a major compound (62%) in the oil of *O. gratissimum* indigenous to Nigeria but cultivated and collected in Taiwan. Nigerian *O. gratissimum* oil is thought to contain eugenol and thymol. Sofowora (1970) used the cultivated material of *O. gratissimum*, grown in the garden of the University of Ife, Nigeria. He detected at least 50% thymol in all the samples. Leaves from the top of the plants contained on average 1.2% of volatile oil. El-Said *et al.* (1967) collected *O. gratissimum* specimens from the area around Ibadan, Nigeria. The steam distilled oil contained thymol as the main compound. No eugenol could be found from any of the oils tested. Ntezurubanza *et al.* (1987) reported the chemical composition of the essential oil of *O. gratissimum* grown in Rwanda. They found that the thymol type oil may also contain eugenol (10.7%), the other major compounds were p-cymene (18.3%) and γ-terpinene (22.9%).

Lawrence (1987) summarized the literature data on *O. gratissimum* oil and published the chemical composition of an oil from plants grown in North Africa. In this oil, eugenol was as the main compound (85.1%). The content of α-copaene (1.9%), terpinen-4-ol (2.6%), caryophyllene (3.6%), germacrene-D (2.6%) and α-terpineol (1.2%) was more than one percent, the rest of the identified compounds were present only in trace amounts.

Maia *et al.* (1988) found a new type of the essential oil of *O. gratissimum*. The sample of Brazilian origin contained p-cymene (29.7%) as the main component. The other major components in this oil were (*E*)-β-farnesene (19.0%) and thymol (13.1%).

Fun and Baerheim Svendsen (1990) examined the oil *O. gratissimum* grown in Aruba (Dutch Antilles). The oil yield varied from 0.7 to 1.7% (v/w). Analysis of the essential oil showed that half of the material belonged to the methyl cinnamate chemotype (*trans*-methyl cinnamate 51–64%) and the other half contained sesquiterpene hydrocarbons (6–46% β-caryophyllene, β-elemene 9–12 %) and varying amounts of eugenol (7–20%) as the main volatile components. According to the authors this indicates the existence of a sesquiterpene hydrocarbon-eugenol chemotype of *O. gratissimum*. The *trans*-methyl cinnamate chemotype was reported for the first time in this paper.

CHEMOTYPES OF OTHER *OCIMUM* SPECIES

The chemical composition of the essential oils of other *Ocimum* species except for *O. basilicum*, *O. canum* and *O. gratissimum* has not been studied very intensively. This means that the chemical diversity at the level of terpenoids and phenylpropanes is not clear cut among other *Ocimum* species. However, some older data exists about the chemical composition of the essential oils and chemotypes of *O. kilimandscharicum*, *O. sanctum* and *O. menthaefolium*. Apart from the *Ocimum* species mentioned above some data concerning the main compounds and other major constituents of basil oils are discussed earlier in this chapter.

O. kilimandscharicum

The essential oil of *O. kilimandscharicum* Gürke is rich in camphor (50–70%), *O. pilosum* Roxb. contains citral (34%) and citronellal (41%) and the oil of *O. viride* Willd. may contain thymol from 40 to 65% (Hegnauer 1966).

O. sanctum

The oil *O. sanctum* L. can be classified into four chemotypes as follows: (a) the citral-type, which contains citral as main compound (about 70%), (b) the eugenol-type, which consists of eugenol (about 70%) and methyl eugenol (c. 20%), (c) the methyl chavicol-type, which is rich in methyl chavicol, linalool and cineole and (d) the chavibetonol-type, which consist of chavibetonol as main compound and also eugenol (Hegnauer 1966).

O. menthaefolium

The oils of *O. menthaefolium* Hochts can be divided into four varieties:

(a) *O. menthaefolium* var. *camphorata*: the oil consists of methyl chavicol (40%), camphor (23%), cineole (13%) and α-pinene (10%)
(b) *O. menthaefolium* var. *estragolata*: methyl chavicol as the main compound (73%), linalool (9%), anethole (5%) and traces of limonene

(c) *O. menthaefolium* var. *anisata*: anethole (39%) and methyl chavicol (31%) as the major compounds in addition to limonene (10%) and linalool (9%)
(d) *O. menthaefolium* var. *citratata*: citral (56%) and methyl chavicol (20%) as major compounds, limonene (9%) and geraniol (10%) as minor constituents (Hegnauer 1966).

VARIATION IN OIL COMPOSITION DUE TO ENVIRONMENTAL FACTORS

The variation in essential oil composition of the above described *Ocimum* species, particularly the variation between chemotypes is genetic. The terpene as well as the phenylpropanoid composition in the essential oils of plants is generally under strict genetic control. In addition there are many other factors which affect the chemical composition of the oils. These factors are the climatic conditions, the locality of a plant, fertility, stage of development, harvesting time, isolation technique and storing. The effects of harvesting time and phenological stage at harvest time and the effects of fertilizers on the essential oil yield and the chemical composition of the oils are briefly discussed below. The other general effects on the composition and yield of essential oils are largely described in the literature concerned with essential oils of *Ocimum* and numerous other plant species.

Phenological Stage

Gulati *et al.* (1977) studied the volatile oil yield of a French basil grown in Uttar Pradesh in India. The components identified by this group were: linalool (54.7%), methyl chavicol (33.9%), α-pinene, β-pinene and 1,8-cineole (1.8%). They found that the oil yield was significantly higher if the blossoms were collected three times at intervals of 15 days and the whole plant at the fourth harvest. This method yielded 40 kg essential oil per ha.

In 1981 Fleisher investigated the yield, quality and composition of the essential oil from sweet basil plants (known as true or French basil) grown in Israel (Newe Ya´ar Experiment Station). Linalool, methyl chavicol, 1,8-cineole and eugenol were found to be the main components of the oil. Fleisher reported two chemotypes; linalool–methyl chavicol (A) and linalool–eugenol (B). It has been found that at the same phenological stage the content of essential oil in the plant increases towards the autumn. Fleisher also showed that the oil yield increases from 0.13% at the first vegetative stage in June to a maximum of 0.35% at the third full flowering stage after two earlier harvests in October.

The chemical composition of basil oil varied considerably between the beginning and the end of the bloom period as shown in Table 4.14. The content of linalool increased towards the end of the bloom stage in both chemotypes and decreased at the early seed stage. The concentration of methyl chavicol decreased from early blooming to the end of blooming and increased at the early stage of seed development. Eugenol, instead, increased in one and decreased in another chemotype during blooming. In the distillation process 25–50% of the essential oil remained in the

Table 4.14 The essential oil composition of linalool–methylchavicol (A) and linalool–eugenol (B) types of *Ocimum basilicum* L. at different phenological stages (Fleisher 1981)

Chemotype	*Phenological Stage*	*Linalool (%)*	*Methyl chavicol (%)*	*Cineole (%)*	*Eugenol (%)*
A	Early blooming	37.5	42.9	7.6	4.0
	Middle blooming	40.5	28.9	8.5	10.2
	End of blooming	47.0	19.7	6.6	11.1
	Early seed stage	42.0	24.4	6.5	7.5
B	Early blooming	49.7	5.3	8.8	27.9
	Middle blooming	51.7	2.1	7.3	24.2
	End of blooming	55.4	1.5	4.5	19.5
	Early seed stage	50.6	2.8	4.8	22.0

Chemotypes: (A) linalool-methyl chavicol
(B) linalool-eugenol

water phase after the volatile oils were decanted off. The oil, which remained in the water phase contained proportionally more polar compounds than the decanted oil (Fleischer and Fleischer 1992).

Sofowora (1970) examined the variation of the volatile oil content in the leaves taken from different locations on an *O. basilicum* plant of the eugenol type. He reported that more oil (1.2%) was produced in the young leaves at the top of the plant than in the older ones from the middle (0.6%) and the base of the plant (0.4%).

Kartnig and Simon (1986) investigated the yield and chemical composition of the volatile oil of various cultivars of *O. basilicum* depending on the harvest time. From seven varieties of basil cultivated under identical conditions the yield and composition of the volatile oil were investigated at different harvest times. The highest yield of volatile oil differed considerably and were reached at various periods as shown in Table 4.15. In all the varieties and for all the harvest times, linalool was the main component, ranging from 39.8% to 75.5%, depending on the cultivar. The authors also concluded that no relation could be found between the highest yield of volatile oil and the percentage of quality determining compounds.

Of the environmental conditions water stress is a known factor which induces biochemical, physiological and developmental alterations in the plants. Simon *et al.* (1992) examined the relationship between water stress and essential oil accumulation in *O. basilicum.* In this study the oil yield varied from 0.3% in the control plants to 0.6% in the plants grown under either mild or moderate water stress. Water stress also altered the chemical composition of the essential oil. The content of linalool and methyl chavicol in the leaves increased from 0.8 to 1.0 $\mu g\ g^{-1}$ (d.w.) and from 0.9 to 4.0 $\mu g\ g^{-1}$ (d.w.), respectively and the content of total sesquiterpenes decreased from 1.0 to 0.7 $\mu g\ g^{-1}$ (d.w.). Also the relative percentage of linalool, methyl chavicol, eugenol and sesquiterpenes in the essential oil were altered. Compared to the controls the percentage of linalool decreased from 25.9% to 16.3% in plants grown in

Table 4.15 Essential oil yield (%, v/w) at different harvest times (Kartnig and Simon 1986)

Variety	*Harvest**			
	1.	*2.*	*3.*	*4.*
"Grosses Grünes"	0.40	0.54	0.40	0.20
"Feines"	0.06	0.16	0.42	0.08
"Sweet basil"	0.58	0.40	0.54	0.04
"Feines Grünes"	0.30	0.34	0.34	0.08
"Grossblättriges"	0.48	0.42	0.40	0.06
"Krauses Grünes"	0.58	0.50	0.48	0.22
"Hohes Grünes"	0.60	0.64	0.42	0.38

* Harvesting: 1. One week before blooming, 2. one week after blooming, 3. four weeks after blooming, 4. seven weeks after blooming.

moderate stress conditions. Similarly also eugenol decreased from 4.3% to 0.1% and the sesquiterpenes from 32.5% to 11.8%, respectively. Methyl chavicol, instead, increased from 30.3% in the controls to 49.9% in plants grown under mild stress regimes and to 64.5% in plants grown under moderate stress regimes. The result suggests that even a mild water stress can increase the essential oil content and alter the oil composition (Kartnig and Simon 1986).

Fertilizers

The effect of fertilization on the herb yield of basil grown in a field trial at the University of Helsinki (Finland) was studied during the years 1984–1985 (Hälvä 1987, Hälvä and Puukka 1987). Three chemotypes of basil (*O. basilicum*) were reported: 1) methyl chavicol, 2) linalool and 3) linalool–eugenol. The linalool-eugenol chemotype originated from Budakalasz (Hungary). The essential oil composition was not affected by nitrogen fertilization. The total oil yield decreased slightly along with the increase of the applied fertilizer. Nykänen (1989) reported that an increase of the nitrogen fertilizer supply over the range 0 to 160 kg/ha first decreased the essential oil production from 466 mg/kg fresh herb to 157 mg/kg and then increased it to 405 mg/kg when the nitrogen fertilizer supply was 120 kg/ha.

Nykänen (1986) studied the production of total oil and the major components of basil (*O. basilicum*) as a function of supplied nitrogen fertilizer. The highest essential oil content was found in basil grown without any nitrogen fertilizer. The oil production decreased by 66 % in lands supplied with 80 kg N/ha, but was increased again with larger fertilizer (120 and 160 kg N/ha) doses. Along with the total oil production, the yields of methyl chavicol, fenchone, camphor, terpinen-4-ol, anethole, methyl eugenol and δ-cadinol decreased by about 80% in basil supplied with 80 kg N/ha compared to the amounts obtained without applied fertilizer. In contrast, the amount of eugenol formed with the fertilizer dose of 80 kg N/ha was more than 200 times higher than the amount found in basil grown without nitrogen fertilizer. The highest linalool content was found in plants supplied with 160 kg N/ha (Nykänen 1987).

REFERENCES

Akgul, A. (1981) Volatile oil composition of sweet basil (*Ocimum basilicum* L.) cultivated in Turkey. *Nahrung*, **33**, 87–88.

Baritaux, O., Richard, H., Touche, J. and Derbesy, M. (1992) Effects of drying and storage of herbs and spices on the essential oil. Part I. Basil, *Ocimum basilicum* L. *Flav. Fragr. J.*, **7**, 267–271.

Bernreuther, A. and Schreier, P. (1991) Multidimensional gas chromatography/mass spectrometry: A powerful tool for the direct chiral evaluatioan of aroma compounds in plant tissues. II Linalool in essential oils and fruits. *Phytochem. Analysis*, **2**, 167–170.

Bobin, M.F., Gau, F., Pelletier, J. and Cotte, J. (1991) Etude de L´Arome Basilic. *Rivista Ital. EPPOS*, 3–13.

Boniface, G., Vernin, G. and Metzger, J. (1987) Les diverses techniques d´analyse de donn´es II. Application aux aromes huiles essentielles d basilic. *Parfum. Cosmet. Arom.*, **74**, 75–77.

Brophy, J.J. and Jogia, M.K. (1986) Essential oils from Fijian *Ocimum basilicum* L.. *Flav. Fragr. J.*, **1**, 53–55.

Carmo, M.M., Raposo, E.J., Venâncio, F., Frazao, S. and Seabra, R. (1990) The essential oil of *Ocimum basilicum* L. from Portugal. *J. Ess. Oil Res.*, **2**, 263–264.

Charles, D.J., Simon, J.E. and Wood, K.V. (1990) Essential oil constituents of *Ocimum micranthum* Willd. *J. Agric. Food Chem.*, **38**, 120–122.

Cheng, Y.S. and Liu, M.L. (1983) Oils of *Ocimum basilicum* L; *Ocimum basilicum* var. *minimum* L.; and *Ocimum gratissimum* L. grown in Taiwan. In *Proceedings 9th International Congress of Essential Oils*, Singapore 1983, Essential Oil Technical Papers Book 4 (Publ. 1986).

Conan, J.Y. (1977) Essai de definition d'un label Bourbon pour quelques huiles essentielles de le Réunion. *Rivista Ital. EPPOS*, **59**, 544–549.

Chien, M.J. (1988) A computer data base of essential oils. In B.M. Lawrence, B.D. Morkherjee and B.J. Willis, (eds.), *Flavors and Fragrances: A World Perspective.* Elsevier Science Publishers BV, Amsterdam.

Chogo, J.B. and Crank, G. (1981) Chemical composition and biological activity of the Tanzanian plant *Ocimum suave. J. Nat. Prod.*, **44**, 308–311.

Dro, A.S. (1974) Untersuchungen zur Zusammensetzung und Biogenese des ätherischen Öls von *Ocimum gratissimum* L. PhD thesis, Albert-Ludwigs-Universität, Freiburg, Germany. In B.M. Lawrence (1978) Progress in essential oils. *Perfumer & Flavorist*, **3**, 36–41.

Ekundayo, O., Laakso, I. and Hiltunen, R. (1989) Constituents of the volatile oil from leaves of *Ocimum canum* Sims. *Flav. Fragr. J.*, **4**, 17–18.

Ekundayo, O., Laakso, I., Oguntimein, B., Okogun, J.I., Elujoba, A.A. and Hiltunen, R. (1987) Essential oil composition of two chemodemes of *Ocimum basilicum* from Nigeria. *Acta Pharm. Fenn.*, **96**, 101–106.

El-Said, F., Sofowora, E.A., Malcolm, S.A. and Hofer, A. (1969) An investigation into the efficacy of *Ocimum gratissimum* as used in Nigerian native medicine. *Planta Medica*, **17**, 195–200.

Fatope, M.O. and Takeda, Y. (1988) The constituents of the leaves of *Ocimum basilicum. Planta Medica*, **54**, 190.

Fleisher, A. (1981) Essential oils from two varieties of *Ocimum basilicum* L. Grown in Israel. *J. Sci. Food. Agric.*, **32**, 1119–1122.

Fleisher, Z. and Fleisher, A. (1992) Volatiles of *Ocimum basilicum* traditionally grown in Israel. Aromatic plants of the holy land and the Sinai, Part VIII. *J. Ess. Oil Res.*, **4**, 97–99.

Fun, C.E. and Baerheim Svendsen, A. (1990) Composition of the essential oils of *Ocimum basilicum* var. *canum* Sims and *O. gratissimum* L. grown on Aruba. *Flav. Fragr. J.*, **5**, 173–177.

Gaydou, E.M., Faure, R., Bianchini, J.-P., Lamaty, G., Rakotonirainy, O. and Randriamiharisoa, R. (1989) Sesquiterpene composition of basil oil. Assignment of the ^{1}H and ^{13}C NMR spectra of β-elemene with two-dimensional NMR. *J. Agric. Food. Chem.*, **37**, 1032–1037.

Georgiev, E. and Genov, N. (1977) Gas chromatographic study of macro components in basil oil. Nauch. Tr. Vissh. Inst. Khranit Vkusova Prom. Plovdiv, **20**, 209–217.

Gildemeister, E. and Hoffmann, F. (1961) Die Ätherischen Öle, 4. Auflage, Band VII, Akademie-Verlag, Berlin 1961.

Grayer, R.J., Kite, G.C., Goldstone, F.J., Bryan, S., Paton, A. and Putievsky, E. (1996) Infraspecific taxonomy and essential oil chemotypes in sweet basil, *Ocimum basilicum. Phytochemistry*, **43**, 1033–1039.

Guillaumin, A. (1930) Les Ocimum á Essence, *Bull. Sci. Pharmacol.*, **37**, 431.

Gulati, G.C. (1977) *Ocimum basilicum* Linn. Methyl chavicol type. Paper No. 31, presented at 7th International Essential Oil Congress, Kyoto. In B.M. Lawrence: Progress in Essential Oils, *Perfumer & Flavorist*, **14**, 1989, 45–51.

Gulati, B.C., Duhan, S.P.S., Gupta, R. and Bhattacharya, A.K. (1977a) Die Einführung französischen Basilikumöls (*Ocimum basilicum* L.) in Tarai, Nainital (Uttar Pradesh). *Parfümerie und Kosmetik*, **58**, 165–168.

Gulati, B.C., Shawl, A.S., Garg, S.N., Sobti, S.N. and Pushpangadan, P. (1977b) Essential oil of *O. canum Sims*. (linalool type). *Indian Perfumer*, **21**, 21–25.

Günther, E. (1949) *The Essential Oils*, Vol. **3**, Van Nostrand Company, New York.

Gupta, S.C. (1994) Genetic analysis of some chemotypes in *Ocimum basilicum* var. *glabratum. Plant Breeding*, **112**, 135–140.

Gupta, S.C. and Sobti, S.N. (1990) Inheritance pattern of methyl chavicol and citral in *Ocimum americanum. Indian Perfumer*, **34**, 253–259.

Gupta, S.C. and Sobti, S.N.(1991) Inheritance pattern of linalool and camphor in *Ocimum canum* Sims. *Indian Perfumer*, **35**, 213–217.

Gupta, S. and Sobti, N. (1993) RRL-OC-11, a linalool rich strain of *O. canum* Sims. *Indian Perfumer*, **37**, 261–266.

Gupta, S. and Tawa, A. (1997) Chemical composition of an improved hybrid strain of *Ocimum canum* Sims. (RRL-OC-11). *J. Ess. Oil Res.*, **9**, 375–377.

Hasegawa, Y., Tajima, K., Toi, N. and Sugimura, Y. (1997) Characteristic components found in the essential oil of *Ocimum basilicum* L. *Flav. Fragr. J.*, **12**, 195–200.

Hegnauer, R. (1966) *Chemotaxonomie der Pflanzen*, Band 4, Birkhäuser Verlag, Basel und Stuttgart, Germany.

Holm, Y., Siira, A.-R. and Hiltunen, R. (1989) A chemical study on various basil chemotypes cultivated in Finland. In proceedings of *20th Int. Symposium on Essential Oils*, September 1989, Würzburg, Germany.

Hoppe, H.A. (1975) *Drogenkunde*, Walter de Gruyter, Berlin, Germany.

Hälvä, S. (1987) Studies on fertilization of dill (*Anethum graveolens* L.) and basil (*Ocimum basilicum* L.) III Oil yield of basil affected by fertilization. *J. of Agric. Sci. in Finland*, **59**, 25–29.

Hälvä, S. and Puukka, L. (1987) Studies on fertilization of dill (*Anethum graveolens* L.) and basil (*Ocimum basilicum* L.) I Herb yield of dill and basil affected by fertilization. *J. of Agric. Sci. in Finland*, **59**, 11–17.

Hörhammer, L., Hamidi, A.E. and Richter, G. (1964) Investigation of Egyptian basil oils by simple chromatographic method. *J. Pharm. Sci.*, **53**, 1033–1036.

Janssen, A.M., Scheffer, J.J.C., Ntezurubanza, L. and Baerheim Svendsen, A. (1989) Antimicrobial activities of some *Ocimum* species grown in Rwanda. *J. Ethnopharmacol.*, **26**, 57–63.

Karawya, M.S., Hashim, F.M. and Hifnawy, M.S. (1974) Oils of *Ocimum bacilicum* L. and *Ocimum rubrum* L. grown in Egypt. *J. Agr. Food. Chem.*, **22**, 520–522.

Kartnig, T. and Simon, B. (1986) Gehalt und Zusammensetzung des ätherischen Öles verschiedener Sorten von *Ocimum basilicum* L. in Abhängigkeit vom Erntezeitpunkt. *Gartenbauwissenschaft*, **5**, 223–225.

Kekelidze, N.A. and Beradze, L.V. (1975) The non-phenolic portion of basil essential oil. In B.M. Lawrence (1978) Progress in essential oils. *Perfumer & Flavorist*, **3**, 36–41.

Khanna, R.K., Sharma, O.S., Sharma, M.L., Misra, P.N. and Singh, A. (1988) Essential oil of Clocimum. A strain of *O. gratissimum* L. raised on alkaline soils. *Parfumerie und Kosmetik*, **69**, 564–568.
Kostrzewa, E. and Karwowska, K. (1991) Chemical composition of the essential oil from sweet basil grown in Poland. *Prace Inst. Lab. Bad. Przem Spoz.*, **44**, 97–107.
Laakso, I., Seppänen-Laakso, T., Herrmann-Wolf, B., Kuhnel, N. and Knobloch, K. (1990) Constituents of the essential oil from the Holy basil or Tulsi plant, *Ocimum sanctum. Planta Medica*, **56**, 527.
Lachowicz, K., Jones, G.P., Briggs, D.R., Benvienu, F.E., Palmer, M.V., Mishra, V. and Hunter, M.M. (1997) Characteristics of plants and plant extracts from five varieties of basil (*Ocimum basilicum* L.). *J. Agric. Food Chem.*, **45**, 2660–2665.
Lai, R.N., Sen, T.K. and Nigam, M.C. (1978) Gas Chromatographie des ätherischen Öls von *Ocimum sanctum* L. *Parfümerie und Kosmetik*, **59**, 230–231.
Lang, E. and Hörster, H. (1977) An Zucker gebundene Reguläre Monoterpene. 11. Untersuchungen der Ölbildung und Akkumulation ätherischen Öle in *Ocimum basilicum* Zelkulturen. *Planta Medica*, **31**, 112–118.
Lawrence, B.M., Terhune, S.J. and Hogg, J.W. (1971) Essential oils and their constituents. VI. The so-called "exotic" oil of *Ocimum basilicum* L. *Flavour Ind.*, **2**, 173–176.
Lawrence, B.M., Hogg, J.W., Terhune, S.J. and Pitchifakul, N. (1972) Essential oils and their constituents. IX. The oils of *Ocimum sanctum* and *Ocimum basilicum* from Thailand. *Flavour Ind.*, **3**, 47–49.
Lawrence, B.M. (1978) Progress in essential oils. *Perfumer & Flavorist*, **3**, 36–41.
Lawrence, B.M. (1980) Progress in essential oils. *Perfumer & Flavorist*, **4**, 31–36.
Lawrence, B.M. (1982) Progress in essential oils, *Perfumer & Flavorist*, **7**, 43–48.
Lawrence, B.M. (1986a) Progress in essential oils. *Perfumer & Flavorist*, **11**, 49–52.
Lawrence, B.M. (1986b) A further examination of the variation of *Ocimum basilicum* L. *Proc. 10th Int. Cong. Essential Oils, Fragrances and Flavors*, November 1986, Washington DC, USA.
Lawrence, B.M. (1987) Progress in essential oils. *Perfumer & Flavorist*, **12**, 69–80.
Lawrence, B.M. (1988) A further examination of the variation of *Ocimum basilicum* L. In B.M. Lawrence, B.D. Mookherjee, and B.J. Willis, (eds.), *Flavors and Fragrances: A World Perspective.* Elsevier Science Publishers BV, Amsterdam, pp. 161–170.
Lawrence, B.M. (1989) Progress in essential oils. *Perfumer & Flavorist*, **14**, 45–55.
Lawrence, B.M. (1992a) Progress in essential oils. *Perfumer & Flavorist*, **17**, 45–56.
Lawrence, B.M. (1992b) Chemical components of Labiatae oils and their exploitation. In R.M. Harley and T. Reynolds, (eds.), *Adcances in Labiatae Science,* Royal Botanic Gardens, Kew, UK, pp. 399–436.
Lawrence, B.M. (1992c) Labiatae oils – mother nature´s chemical factory. In B.M. Lawrence, (ed.), *Essential Oils* 1988–1991, Allured Publ, Carol Stream, IL, pp. 188–206.
Lawrence, B.M. (1995) Progress in essential oils. *Perfumer & Flavorist*, **20**, 29–41.
Lawrence, B.M. (1997) Progress in essential oils. *Perfumer & Flavorist,* **22**, 57–74.
Maia, J.G.S., Ramos, L.S., Luz, A.I.R., da Silva, M.L., Zoghbi, B. and das Gracos, M. (1988) Uncommon Brazilian essential oils of the Labiatae and Compositae. In B.M. Lawrence (1997), Progress in essential oils, *Perfumer and Flavorist*, **22**, 57–74.
Manninen, P., Riekkola, M.L., Holm, Y. and Hiltunen, R. (1990) SFC in analysis of aromatic plants. *J. High Res. Chromatog.*, **13**, 167–169.
Mariani, R., Ducci, L., Scotti, A. and Gravina, A. (1991) Lútilaisation de léspace de tete dynamique dans la selection des basilics. *Rivista Ital. EPPOS* (Special Issue), 125–137.
Masade, Y. (1976) *Analysis of essential oils by gas chromatography and mass spectrometry*, John Wiley & Sons, New York.
De Medici, D., Pieretti, S., Salvatore, G., Nicoletti, M. and Rasoanaivo, P. (1992) Chemical analysis of essential oils of Malagasy medicinal plants by gas chromatography and NMR spectroscopy. *Flav. Fragr. J.*, **7**, 275–281.

Modawi, B.M., Duprey, R.J.M., El Magboul, A.Z.I. and Satti, A.M. (1983) Constituents of the essential oil of *O. basilicum* var. *thyrsiflorum*. *Fitoterapia*, **LV**, 60–62.

Mwangi, J.W., Lwande, W. and Hassanali, A. (1994) Composition of the leaf oil of *Ocimum keniense* Auobangira. *Flav. Fragr. J.*, **9**, 75–76.

Ntezurubanza, L., Scheffer, J.J.C., Looman, A. and Baerheim Svendsen, A. (1984) Composition of essential oil of *Ocimum kilimandscharicum* grown in Rwanda. *Planta Medica*, **50**, 385–387.

Ntezurubanza, L., Scheffer, J.J.C. and Looman, A. (1985) Composition of the essential oil of *Ocimum canum* grown in Rwanda. *Pharmaceutisch Weekblad, Scientific Edition*, **7**, 273–276.

Ntezurubanza, L., Scheffer, J.J.C. and Baerheim Svendsen, A. (1986) Composition of the essential oil of *Ocimum trichodon* grown in Rwanda. *J. Nat. Prod.*, **49**, 945–947.

Ntezurubanza, L., Scheffer, J.J.C. and Baerheim Svendsen, A. (1987) Composition of the essential oil of *Ocimum gratissimum* grown in Rwanda. *Planta Medica*, **53**, 421–423.

Nykänen, I. (1986) High resolution gas chromatographic-mass spectrometric determination of flavour composition of basil (*Ocimum basilicum* L.) cultivated in Finland. *Z. Lebensm. Forsch.*, **183**, 172–176.

Nykänen, L. and Nykänen, I. (1987) The effect of drying on the composition of the essential oil of some Labiatae herbs cultivated in Finland. In M. Martens, G.A. Dalen and H. Russwurm (eds.), *Flavour Science and Technology*, John Wiley & Sons, New York, pp. 83–88.

Nykänen, I. (1987) *Gas chromatographic and mass spectrometric investigation of the flavour composition of some Labiatae herbs cultivated in Finland.* PhD thesis, University of Helsinki, Finland.

Pérez-Alonso, M.J., Velasco-Negueruela, A., Emin Duru, M., Harmandar, M. and Esteban, J.L. (1995) Composition of the essential oils of *Ocimum basilicum* var. *glabratum* and *Rosmarinus officinalis* from Turkey. J. *Ess. Oil Res.*, **7**, 73–75.

Peter, H.H. and Rémy, M. (1978) L'essence de basilic, étude comparative. *Parfum, Cosmét., Arômes*, **21**, 61–68.

Pino, J.A., Rosado, A. and Fuents, V. (1996) Composition of the essential oil from the leaves and flowers of *Ocimum gratissimum* L. grown in Cuba. *J. Ess. Oil Res.*, **8**, 139–141.

Pluhar, Z., Nemeth, E. and Gyogynovenytermesztesi, T. (1996) Analysis of supercritical extracts from essential oils. (Abstract, in Hungarian). Olaj Szappan, Kosmet. **45**, 70–74.

Pogany, D. (1967) *Composition of oil of sweet basil (Ocimum basilicum L.) obtained from plants grown at different temperatures.* PhD thesis, University of Illinois, Chicago.

Prasad, G., Kumar, A., Singh, A.K., Bhattacharya, A.K. and Singh, K. (1986) Antimicrobial activity of essential oils of some *Ocimum* species and clove oil. *Fitoterapia*, **LVIII**, 429–432.

Pushpangadan, P. and Sobti, S. (1982) Cytogenetical studies in the genus *Ocimum*. I Origin of *O. americanum*, cytotaxonomival and experimental proof. *Cytologia*, **47**, 575–583.

Pushpangadan, P. and Bradu, B.L. (1995) Medicinal and Aromatic Plants. In K.L.Chadha and R. Gupta, (eds.), *Advances in Horticulture*, **11**, Malhotra Publishing House, New Delhi, pp. 627–657.

Randriamiharisoa, R. and Gaydou, E.M. (1985) Relation entre la teneur en linalool at le pouvoir rotatoire dans les essences de basilic. *Parfum. Cosmet. Arom.*, **64**, 89–90.

Randriamiharisoa, R., Gaydou, E.M., Bianchini, J.P., Ravelojoana, G. and Vernin, G. (1986) Etude de la variation de la composition chimique et classification des huiles essentielles de Basilic de Madagascar. *Sci. Aliment.*, **6**, 221–231.,

Reverchon, E., Osseo, L.S. and Gorgoglione, D. (1994) Supercritical CO_2 extraction of basil oil: characterization of products and process modelling, *J. Supercritical Fluids*, **7**, 185–190.

Roque, R. (1991) Composicao do olea essencial de *Ocimum basilicum* L. cultivado. *Bol. Fac. farm. Coimbra*, **15**, 47–51.

Rovesti, P. (1952) Ind. Parfumerie, **7**, 165. In R. Hegnauer, *Chemotaxonomie der Planzen, Band* **4**, Birkhäuser Verlag, Basel und Stuttgart, pp. 313–316 (1966).

Rovesti, P. (1952) Rivista Ital. Essenze, Porfumi, **34**, 247. In R. Hegnauer, *Chemotaxonomie der Planzen, Band* 4, Birkhäuser Verlag Basel und Stuttgart, pp. 313–316 (1966).

Ruberto, G., Spadaro, A., Piattelli, M., Piozzi, F. and Passannanti, S. (1991) Volatile flavour components of *Ocimum basilicum* var. *hispidum* (Lam.) Chiov. *Flav. Fragr. J.*, **6**, 225–227.

Sainsbury, M. and Sofowora, E.A. (1971) Essential oil from leaves and inflorescences of *Ocimum gratissimum. Phytochemistry*, **10**, 3309–3310.

Sheen, Y.L., Tsai Ou, Y.H. and Tsai, S.J. (1991) Flavor characteristic compounds found in the essential oil of *Ocimum basilicum* L. with sensory evaluation and statistical analysis. *J. Agric. Food Chem.*, **39**, 939–943.

Schreier, P. (1984) Chromatographic studies of biogenesis of plant volatiles. In W. Bersch, W.G. Jennings and R.E. Kaiser, (eds.), *Chromatographic Methods*, Hüthig Verlag, Heidelberg-Basel-New York, pp. 84–126.

Simon, J.E., Quinn, J. and Murray, R. G. (1990) Basil: A Source of Essential Oils. In J. Janick and J.E. Simon, (eds), *Advances in New Crops* (Proceedings of the First National Symposium NEW CROPS: Research, Development, Economics, Indianapolis, Indiana, October, 1988), Timber Press, Portland, Oregon, pp. 484–489.

Simon, J.E., Reiss-Bubenheim, D., Joly, R.J. and Charles, D.J. (1992) Water-stress induced alterations in essential oil content and composition of sweet basil. *J. Ess. Oil Res.*, **4**, 71–75.

Skrubis, B. and Markakis, P. (1976) The effect of photo-periodism on the growth and the essential oil of *Ocimum basilicum* (sweet basil). *Econ. Baran.*, **30**, 389–393.

Sobti, S.N., Pushpangadan, P. and Atal, C.K. (1976) Genus *Ocimum* – a potential source of new essential oils. *Indian Perfumer*, **20**, 59–68.

Sobti, S.N. and Pushpangdan, P. (1982) Cultivation and Utilization of Aromatic Plants., Regional Reseach Laboratory, Jammu Tawi, pp. 457–472. In Khanna, R.K., Sharma, O.S., Sharma, M.L., Misra, P.N. and Singh, A. (1988) Essential Oil of Clocimum. A Strain of *O. gratissimum* L. raised on Alkaline Soils. *Parfumerie und Kosmetik*, **69**, 564–568.

Sobti, S.N., Pushpangdan, P. and Atal, S.N. (1982) Cultivation and Utilization of Aromatic Plants., Regional Reseach Laboratory, Jammu Tawi, pp. 473–486. In Khanna, R.K., Sharma, O.S., Sharma, M.L., Misra, P.N. and Singh, A. (1988) Essential Oil of Clocimum. A Strain of *O. gratissimum* L. raised on Alkaline Soils. *Parfumerie und Kosmetik*, **69**, 564–568.

Sofowora, E.A. (1970) A Study of the variations in essential Oil of cultivated *Ocimum gratissimum. Planta Medica*, **18**, 173–176.

Srinivas, S.R. (1986) Atlas of Essential Oils. In B.M. Lawrence (1989), Progress in Essential Oils, *Perfumer & Flavorist*, **14**,. 45–52.

Staath, N.A. and Azzo, N.R. (1993) In. G. Charalambous, (ed.), *Food Flavors, Ingredients and Composition*, Elsevier Sci. Publ. B.V., Amsterdam, pp. 591–603.

Suchorska, K. and Osinska, E. (1992) Variability of the purple and the lettuce-leaved sweet basil (*Ocimum basilicum* L.). *Herba Polonica*, **XXXVIII**, 175–181.

Tapanes, R., Delgado, M. and Correa, M.T. (1985) Obtenciony analisis mediante la GC-EM de un aceite esencial de *Ocimum basilicum* L. Rev. Cienc. Quim., **16**, 217–220.

Tateo, F., Santamaria, L., Bianchi, L. and Branchi, A. (1989) Basil oil and tarragon oil: composition and genotoxicity evaluation. *J. Ess. Oil Res.*, **1**, 111–118.

Terhune, S.J., Hogg, J.W. and Lawrence, B.M. (1974) Bicyclosesquiphellandrene and 1-epibicyclosesquiphellandrene: two new dienes based on the cadalene skeleton. *Phytochemistry*, **13**, 1183–1185.

Thomas, O.O. (1989) Re-examination of the antimicrobial activities of *Xylopia aethiopica, Carica papaya, Ocimum gratisssimum* and *Jatropa curcas*. Fitoterapia, **LX**, 147–155.

Tsai, S.J. and Sheen, L.Y. (1987) Essential oil of *Ocimum basilicum* L. cultivated in Taiwan. In L.W. Sze and F.C. Woo, (eds.), *Trends in Food Science*, Proceedings of the 7th World Congress of Food Science and Technology, Singapore Institute of Food Science and Technology, Singapore, pp. 66–70.

Upadhyay, R.K., Misra, L.N. and Singh, G. (1991) Sesquiterpene alcohols of the copaene series from essential oil of *Ocimum americanum. Phytochemistry*, **30**, 691–693.

Vernin, G., Metzger, J., Fraisse, D. and Scharff, D. (1984) Analysis of basil oils by GC-MS data bank. *Perfumer & Flavorist*, **9**, 71–86.

Vostrowsky, O., Grabe, W. and Maia, J.G.S. (1990) Essential oil of Alfavaca, *Ocimum gratissimum* from Brazilian Amazon. *Zeitschr. Naturforsch*, **45c**, 1073–1076.

Wu, M.-Z., Xiao, S.-C., Ren, W.-J. and Chem, P.-Q. (1990) Chemical components of the essential oil from an escape *Ocimum gratissimum* L. var. *suave* Willd. in Miyi county, Sichuan. *Tianran Chanwu Yanjiu Yu Kaifa*, **2**, 58–60.

Xaasan, C.C. and Cardulraxmaan, A.D. (1981) Constituents of the essential oil of *Ocimum canum*. *J. Nat. Prod.*, **44**, 752–753.

Xaasan, C.C., Cabdulraxmaan, A.D., Passannanti, S., Piozzi, F. and Schmid, J.P. (1981) Constituents of the essential oil of *Ocimum canum*. *Lloydia*, **44**, 752–753

Zamureenko, V.A., Klyuev, N.A., Dmitriev, L.B., Polakova, S.G. and Grandberg, I.I. (1986) Composition of the essential oil in eugenol-type basils. *Izv. Timiryazevsk S-kh. Akad.*, **2**, 172–175.

Zola, A. and Garnero, J. (1973) Contribution a l´etude de quelques essences de basilic de typ europeen. *Parfum. Cosmet. Savon. France*, **3**, 15–19.

Yu, X.J. and Cheng, B.Q. (1986) Analysis of the chemical constituents of *Ocimum gratissimum* var. *suave* oil. *Yunnan Zhiwu Yanjiu*, **8**, 171–174,

Özek, T., Beis, S.H., Demircakmak, B. and Baser, K.H.C. (1995) Composition of the essential oil of *Ocimum basilicum* L. cultivated in Turkey. *J. Ess. Oil Res.*, **7**, 203–205.

5. BIOACTIVITY OF BASIL

YVONNE HOLM

Department of Pharmacy, P.O Box 56, FIN-00014
University of Helsinki, Finland

TRADITIONAL MEDICINE

Basil has traditionally been used for head colds and as a cure for warts and worms, as an appetite stimulant, carminative, and diuretic. In addition, it has been used as a mouth wash and adstringent to cure inflammations in the mouth and throat. Alcoholic extracts of basil have been used in creams to treat slowly healing wounds (Wichtl 1989). Basil is more widely used as a medicinal herb in the Far East, especially in China and India. It was first described in a major Chinese herbal around A.D. 1060 and has since been used in China for spasms of the stomach and kidney ailments, among others. It is especially recommended for use before and after parturition to promote blood circulation. The whole herb is also used to treat snakebite and insect bites (Leung and Foster 1996).

In Nigeria, a decoction of the leaves of *O. gratissimum* is used in the treatment of fever, as a diaphoretic and also as a stomachic and laxative. In Franchophone West Africa, the plant is used in treating coughs and fevers and as an anthelmintic. In areas around Ibadan (Western State of Nigeria), *O. gratissimum* is most often taken as a decoction of the whole herb (Agbo) and is particularly used in treating diarrhoea (El-Said *et al.*, 1969). It is known to the Yorubas as "Efirin-nla" and to the Ibos as "Nehanwu", in Nigeria (Thomas 1989).

In Kenya most members of *Ocimum* are used as expectorants. The people in rural areas are the principle utilizers of the herbal remedies.

Ocimum basilicum is known under the following local names in the Rift Valley in central Kenya (area in brackets): Chemishwa (Tugen), Chenekom/Sipko (Pokot), Embuke/Emboa (Bukusu), Lemurran (Samburu), Mwenye (Luhya), Mutaa (Kamba) and Rigorio (Marakwet). The vapour of boiling leaves is inhaled for nasal or bronchial catarrh and colds. The leaves may be rubbed between the palms and sniffed for colds. It cures stomach-ache and constipation. The leaves are crushed and the juice is used as vermifuge. It is further used to repel mosquitoes and as a broom to sweep chicken house in order to get rid of fleas.

Ocimum gratissimum is known under the name Mutaa (Kamba). The leaves are eaten for stomach-ache. Pounded leaves are soaked in water and concoction is used as insecticide on maize cobs. *Ocimum kilimandscharicum* is known under the names Gethereti/Makori (Meru), Lisuranza/Mwonyi (Luhya), Mbirirwa (Marakwet) and Mutei (Kikuyu). The leaves treat congested chest, cough and cold, by sniffing crushed leaves or inhaling vapour of boiling leaves. Infusion is a cure for measles. It is also used to repel insects.

Ocimum pseudokilimandscharicum is known by the name Mukandu munini in the Kamba area. The leaves are ground to powder, mixed with water and drunk to cure stomach-ache.

Ocimum suave is known under the names Chemwoken (Pokot), Mukandu (Kamba), Makanda kandu (Meru), Yoiyoiya/Chesimia (Marakwet), Mkandu (Kikuyu), Sivai (Kipsigis) and Sunoni (Masai). The leaves are rubbed between palms and sniffed to treat a blocked nose and cough. The Meru prepare an infusion of the leaves for flu. Chewed leaves treat toothache, the juice is an anthelmintic in small children and roots boiled in soup treats chestache. It may be burnt at night to drive away insects (Githinji and Kokwaro 1993). In Tanzania *O. suave* have been claimed to have various medicinal activities, and extracts of the plant are used for treating coughs, eye and ear complaints, and abdominal pains (Chogo and Crank 1981).

O. canum, *O. gratissimum*, *O. trichodon* and *O. urticifolium* (*O. suave*) are used in Rwanda as infusions and for inhalation of their aromatic vapours, as sneezing powders for curing headache and madness (Janssen *et al.*, 1989).

In Congo *O. gratissimum* aerial parts are taken as febrifuge, against cough and angin (Ndounga and Ouamba 1997).

Ocimum sanctum is a great sacred medicinal plant in India. It is called Thulasi (or Vishnupriya) in Sanskrit, Thulasi in Tamil and Kala-Thulasi in Hindi. It is used for the prevention of pregnancy among the Irulars, the tribal of Anaikkatty Hills, Coimbatore, Tamil Nadu. Roots of *Piper betle* and half the quantity of fruits of *Piper nigrum* are made into a paste in leaf juice of *O. sanctum*. It is administered internally, 5 g/day, for both male and female for 96 days to get complete sterility (Lakshmanan and Sankara Narayanan 1990).

O. basilicum (Albahaca) is one of many plants used in Guatemala to treat gastrointestinal disorders such as colic, stomach pains, intestinal parasites, flatulence, and loss of appetite. It is also used as an anti-emetic agent (Caceres *et al.*, 1990). Among the Caribs, a small ethnic group of Afro-Caribbean origin, in Guatemala a decoction of *O. micranthum* (Albahaca) leaves is used orally for coughs and phlegm, stomach pains, intestinal parasites and skin diseases (locally). The juice is used directly for ear pain (Girón *et al.*, 1991).

MODERN USE

Basil is used as a fragrance ingredient in perfumes, soaps, hair dressings, dental creams, and mouth washes. The most extensive use is as a spice in all major food products, usually in rather low levels (mostly below 0.005%). The use in foodstuffs is discussed in another chapter in this book. The fresh herb is considered by some a source of vitamin C (Leung and Foster 1996). Basil is subject of a German therapeutic monograph as stomachicum (Bundesanzeiger 1992).

ANTIMICROBIAL ACTIVITY

Antibacterial Activity

Aqueous extracts or infusions of *O. gratissimum* showed no activity against the test organisms (*Salmonella* spp., *Shigella sonnei*, *Shigella schmitzi*, *Staphylococcus aureus* and

Escherichia coli). This indicates that the antibacterial principles are not watersoluble. A high phenol content, in this case thymol, of the oil gives a higher antibacterial activity (Thomas 1989, El-Said *et al.*, 1969). Saturated aqueous solutions of *O. gratissimum* oil and of thymol, respectively, showed inhibitory effects on the growth of *Salmonella* spp., *E. coli*, *Staphylococcus aureus*, *Shigella sonnei* and *S. schmitzi*. The minimum inhibitory concentration for the oil was 20.0–27.5% and for thymol 11.0–20.0% depending on the bacteria in question. Thymol was more active than the oil of *O.gratissimum*, but neither had any appreciable activity against *Pseudomonas aeruginosa* (MIC > 75%) The aqueous solutions of thymol and *O. gratissimum* oil were most effective against *S. schmitzi* (El-Said *et al.*, 1969).

The antibacterial effect of *O. gratissimum* oil is frequently reported (El-Said *et al.*, 1969, Grover and Rao 1977, Ramanoelina *et al.*, 1987, Janssen *et al.*, 1989, Thomas 1989, Jedlickova *et al.*, 1992, Ilori *et al.*, 1996, Ndounga and Ouamba, 1997). El-Said *et al.* (1969) found that *O. gratissimum* oil was active against *Escherichia coli*, *Klebsiella aerogenes*, *Proteus* spp., *Salmonella* spp., *Shigella schmitzi*, *Shigella sonnei*, *Bacillus subtilis*, *Sarcina lutea* and *Staphylococcus aureus*. The oil showed no activity against *Pseudomonas aeruginosa*, an organism which is known to be resistant to many antibacterial agents. Grover and Rao (1977) found that the oil of *O. gratissimum* was active against a number of bacteria (Table 5.1, superscript 2). The oil was not active against *Klebsiella pneumoniae*. Ramanoelina *et al.* (1987) report that the oil of *O. gratissimum* was tested on eight referied strains of bacteria commonly used for antibiotic measurements and also on twelve other enteropathogenic bacteria strains and showed a large spectrum action. Janssen *et al.* (1989) determined the maximum inhibitory dilution (MID) of *O. gratissimum* oil against *Escherichia coli* and *Staphylococcus aureus*, and the results were 1:1600 and 1:3200, respectively. According to earlier reports (El-Said *et al.*, 1969, Grover and Rao 1977) *O. gratissimum* oil showed no antibacterial activity against *Pseudomonas aeruginosa* and *Klebsiella pneumoniae*, but Ndounga and Ouamba (1997) showed that the oil had a low inhibitory effect on *P. aeruginosa* (+) and a quite good inhibitory effect on *K. pneumoniae* (+++). These opposite results are probably due to different techniques used for the testing and different oil compositions. El-Said *et al.* (1969) used drops of oil on the surface of seeded agar plates, which were incubated and zones of inhibition measured. Grover and Rao (1977) used the disc diffusion method with discs dipped in the essential oil and measured inhibition zones. Ndounga and Ouamba (1997) also used the disc diffusion method. Janssen *et al.* (1989) state that a combination of techniques should be used to obtain optimum information of the antibacterial activity. The biogram technique should be applied to investigate which constituent of an oil might be responsible for an observed activity, and the dilution technique to determine the maximum inhibitory dilution-value, which is a more fundamental parameter of antimicrobial activity than the inhibition diameter obtained by the agar overlay technique. The minimum inhibitory concentration of *O. gratissimum* oil was 312.5 μg/ml for *Staphylococcus aureus*, *Escherichia coli*, *Salmonella* sp., *Serratia marcescens*, *Klebsiella pneumoniae* and *Proteus vulgaris* and 625 μg/ml for *Streptococcus faecalis* and *Pseudomonas aeruginosa* (Ndounga and Ouamba 1997).

The antibacterial effect of *O. suave* (*O. urticifolium*) oil is less investigated. Chogo and Crank (1981) report the following minimum inhibitory concentrations for *O. suave* oil: *Escherichia coli* 900, *Micrococcus luteus* 700 and *Staphylococcus aureus* 800 μg/ml. The same

Table 5.1 Antibacterial activity of *Ocimum* species (+ indicates that the plant is active against the bacterium in question, - that it is not active)

Bacteria	*O. gratissimum*	*O. suave* (*O. urticifolium*)	*O. sanctum*	*O. basilicum*	*O. canum*	*O. trichodon*	*O. kilimandscharicum*
Aeromonas hydrophila	-11						
Aeromonas sobria	+11						
Bacillus anthraces	+2		+2,10	+10			-10
Bacillus saccharolyticus			+10	+10			+10
Bacillus stearothermophilus			+10	+10			+10
Bacillus subtilis	+1,2,5	+5	+2,10	+4,10	+5	+5	+10
Bacillus thurengiensis			+10	+10			+10
Citrobacter sp.			+10	+10			-10
Enterobacter sp.			+10	-10			-10
Escherichia coli	+1,2,5,6,9,11	+3,5	+2, -10	+4,7,9, -10	+5	+5	-10
Klebsiella aerogenes	+1						
Klebsiella pneumoniae	-2, +9		-2	+9			
Lactobacillus casei			+10	+10			+10
Lactobacillus plantarum			+10	+10			+10
Micrococcus glutamicus			+10	+10			+10
Micrococcus luteus		+3					
Plesiomonas shigelloides	+11						
Proteus spp.	+1						
Proteus vulgaris	+2,9		+2	+9			
Pseudomonas aeruginosa	-1,11, +2,9		+2	+4, -9			
Pseudomonas sp.			-10	-10			-10
Salmonella spp.	+9		-10	+9 -10			-10
Salmonella enteritidis				+			
Salmonella newport	+2		+2				
Salmonella pullorum	+2		+2				
Salmonella richmond	+2		+2				
Salmonella saintpaul			-10	-10			-10

Table 5.1 (continued)

Salmonella stanley	+2		+2				
Salmonella typhi	+11				–		
Salmonella typhimurium	+2		+2				
Salmonella weltevreden			+10	+10			–10
Sarcina lutea	+1		+10	+10			+10
Shigella dysenteriae	+11				–7		
Shigella flexneri				+7			
Shigella schmitzi	+1						
Shigella sonnei	+1						
Staphylococcus aureus	+1,2,5,6,9	+3,5	+2	+4,8,9,10	+5	+5	+10
Staphylococcus sp.			+10	+10			+10
Serratia marcescens	+9			+9			
Streptococcus faecalis	+9			–9			

[1] El-Said *et al.*, 1969
[2] Grover and Rao 1977
[3] Chogo and Crank 1981
[4] Janssen *et al.*, 1986
[5] Janssen *et al.*, 1989
[6] Thomas 1989
[7] Caceres *et al.*, 1990
[8] Abdel-Sattar *et al.*, 1995
[9] Ndounga and Ouamba 1997
[10] Prasad *et al.*, 1986
[11] Ilori *et al.*, 1996

values for eugenol (main component of *O. suave* oil) are 600, 600 and 700 μg/ml, respectively. The authors also tested *O. suave* oil from which the eugenol was removed and got inhibitory concentrations > 1000 μg/ml, which according to them indicate that eugenol is the active substance. Janssen *et al.* (1989) report maximum inhibitory dilutions for *O. suave* oil against *E. coli* and *S. aureus*. For *E. coli* the values are 1:400 or 1:800 and for *S. aureus* 1:400, 1:800 or 1:1600 depending on the oil sample. They also present a growth curve of *S. aureus* obtained for dilutions of the essential oil of *O. suave*, which indicate that mainly the lag times changed and that the growth rates in the logarithmic phase were little influenced by the oil dilutions.

The oil of *O. sanctum* is surprisingly little investigated with respect to the antibacterial activity considering its extensive use and the research going on in other areas. The essential oil of *O. sanctum* was active against a number of bacteria (Table 5.1, superscripts 2 and 10), but it showed no activity against *Klebsiella pneumoniae.* Prasad *et al.* (1986) found that *O. sanctum* oil was not active against *Salmonella saintpaul*, *Salmonella* spp. and *Pseudomonas* sp. Compared to the oil of *O. gratissimum* it seemed to have a higher antibacterial activity. Even 0.2% dilutions of the oil were active against most of the tested bacteria, except for *Pseudomonas aeruginosa* (Grover and Rao 1977).

O. basilicum oil is antibacterial against *Escherichia coli*, *Pseudomonas aeruginosa*, *Bacillus subtilis* and *Staphylococcus aureus*. The highest inhibition diameter, 17.7 mm, was measured with *B. subtilis* (Janssen *et al.*, 1986). *O. basilicum* oil was mildly active against *S. aureus*, *E. coli*, *Salmonella* sp., *Serratia marcescens*, *Klebsiella pneumoniae* and *Proteus vulgaris*. It showed no activity against *Streptococcus faecalis* and *Pseudomonas aeruginosa.* Minimum inhibitory concentrations measured varied from 1250–5000 μg/ml (Ndounga and Ouamba 1997). The antibacterial activity of *O. basilicum* oil varied according to its origin, i.e. probably due to different composition of the oil. For instance French *O. basilicum* oil was not active against *Staphylococcus aureus* while both the Indian and Niazbo *O. basilicum* oils were active. In general *O. basilicum* oil seemed to have a higher activity against gram positive than against gram negative bacteria (Prasad *et al.*, 1986). A chloroform extract of *O. basilicum* leaves showed antibacterial activity against *S. aureus* (Abdel-Sattar *et al.*, 1995). A 50% ethanol extract of *O. basilicum* leaves showed antibacterial activity against *E. coli*, *Salmonella enteritidis* and *Shigella flexneri* (Caceres *et al.*, 1990).

Essential oils of *O. canum* and *O. trichodon* showed antibacterial activity against *E. coli*, *B. subtilis* and *S. aureus.* Maximum inhibitory dilution for *O. canum* oil was 1:400 for *E. coli* and 1:800 for *S. aureus*. Corresponding values for *O. trichodon* oil were 1:400 and 1:1600, respectively (Janssen *et al.*, 1989). The essential oil of O. *kilimandscharicum* was active against a number of bacteria (Table 5.1, superscript 10), all of them gram positive. No effect was found against the tested gram negative bacteria (Prasad *et al.*, 1986).

Antifungal Activity

The antifungal activity of *Ocimum* leaves, extracts, essential oils and their components is frequently studied, mostly in warm countries where the need for protection of plants and stored crops against fungi is of great importance. Also the effect of *Ocimum* oils against a number of dermatophytes has been studied. Data concerning *Ocimum* species and their antifungal activity have been collected in Table 5.2.

Table 5.2 Antifungal activity of *Ocimum* species (+ active, - not active against the fungus in question)

Fungus	*O. gratissimum*	*O. urticifolium (O. suave)*	*O. basilicum (French)*	*O. basilicum (Indian)*	*O. basilicum (Niazbo)*	*O. sanctum*	*O. kilimandscharicum*	*O. canum*	*O. adscendens*
Alternaria alternata	+12,18					+5		+6	
Alternaria brassicae								+6	
Alternaria humicola								+6	
Alternaria solani						+13		+6	
Alternaria sp.			+2	+2	+2	-2	-2		
Alternaria tenuissima								+6	
Aspergillus awamori			+26						
Aspergillus candidus								+6	
Aspergillus chevalieri			+26						
Aspergillus flavus	-7		+20			+4,11, -7		+22	+21
Aspergillus fumigatus	-7,25, +9		+2,26		+2	+2,9	+2,5,11, -7	+2	
Aspergillus nidulans			+26						
Aspergillus niger	-7		+2,26		+2	+2	-2,5, +7	-2	+6
Aspergillus ochraceus			+26						
Aspergillus parasiticus			+2	+2	+2	+2,11	-2		
Aspergillus ruber			+26						
Aspergillus sulphureus								+6	
Aspergillus sydowi			+26					+6	
Aspergillus tamarii			+26						
Basidiobolus haptosporus	+25								
Basidiobolus ranarum	+25								
Botryodiplodia theobromae			+26			+4			
Candida albicans	+7,9,-25		+2,8	-2	+2,9	-2,7	+2		
Chaetomium indicum			+26						
Cladosporium cladosporioides								+6	
Cladosporium herbareum			+26					+6	
Cochliobolus sativus			+26						
Colletotrichum capsici	+12		+26			+13			
Colletrotrichum sp.								+6	
Cryptococcus neoformans			+2	+2	+2	+2	+2		
Curvularia lunata			+26			+4,16			

Table 5.2 (continued)

Curvularia specifera			+26						
Drechslera auntii								+6	
Epicoccum nigrum			+26					+6	
Epidermophyton floccosum	+23, 24		+3						
Fusarium acuminatum			+26						
Fusarium equiseti			+26						
Fusarium moniliforme			+26						
Fusarium oxysporum			+26			+4			
Fusarium oxysporum									
f. sp. ciceri								+6	
Fusarium semitectum			+26					+6	
Fusarium sesami								+6	
Fusarium solani						+13			
Geotrichum candidum	_25								
Helminthosporium oryzae						+15		+6	+10
Helminthosporium oxysporum						+11			
Helminthosporium spiciferum						+4			
Histoplasma capsulatum			_2	_2	+2	_2	_2		
Macrophomina phaseoli			+26						
Microsporum canis	+23, 24		+2	+2	+2	+2	+2		
Microsporum gypseum			+2	+2	+2	+2	_2		
Mucor mucedo					+5				
Penicillium chrysogenum			+26						
Penicillium citrinum			+26					+6	
Penicillium digitatum	_7					+7			
Penicillium italicum			+19						
Pyricularia oryzae						+15			
Pythium aphanidermatum								+6	
Pythium debaryanum								+6	
Pythium proliferum								+6	
Rhizoctonia solani	+18					+15		+6,14,18	
Rhizopus arrhizus						+5			
Rhizopus stolonifer	_7					+7			
Saccharomyces cerevisae		+17							
Sclerothium rolfsii	+12,18		+26						
Sporotrichum schenkii			+2	_2	+2	+2	+2		

Table 5.2 (continued)

Trichoderma hargianum								+6
Trichoderma viride						+11		
Trichophyton mentagrophytes	+9,23,24,25		+2	+2	+2, 9	+2	+2	
Trichophyton mentagrophytes var. *interdigitale*	+1	+1	+3					
Trichophyton rubrum	+24,25		+2,3	+2	+2	+2	+2	
Trichophyton verrucosum			+2	_2	+2	+2	+2	
Trichothecium roseum							+4	

[1] Janssen *et al.*, 1989
[2] Prasad *et al.*, 1986
[3] Janssen *et al.*, 1988
[4] Singh *et al.*, 1993
[5] Saksena and Tripathi 1985
[6] Pandey and Dubey 1994
[7] Grover and Rao 1977
[8] Janssen *et al.*, 1986
[9] Ndounga and Ouamba 1997
[10] Asthana *et al.*, 1982
[11] Prakash and Singh 1986
[12] Tripathi *et al.*, 1985
[13] Dey and Choudhuri 1984
[14] Pandey and Dubey 1992
[15] Tewari and Mandakini 1991
[16] Upadhyaya and Gupta 1990
[17] Chogo and Crank 1981
[18] Thakur *et al.*, 1989
[19] Arora and Pandey 1977
[20] Awasthi 1991
[21] Asthana *et al.*, 1986
[22] Mishra *et al.*, 1989
[23] Singh *et al.*, 1985
[24] Lima *et al.*, 1993
[25] Nwuosu and Okafor 1995
[26] Dube *et al.*, 1989

Protection of plants and stored crops

An ethanolic extract of *O. sanctum* was used to treat healthy ripe tomato fruits prior to and after inoculation with *Aspergillus niger* in the presence of *Drosophila busckii*. The treatment kept the fruits free from rotting for 5 to 7 days (Sinha and Saxena 1989). The essential oil of *O. canum* was effective against damping-off disease causing fungi, *Pythium aphanidermatum*, *P. debaryanum* and *Rhizoctonia solani*. *O. canum* could control damping-off disease of tomato up to 50% in soil infected with *P. aphanidermatum* and up to 43% in soil infected with *P. debaryanum*. The essential oil was not phytotoxic and showed superiority over commonly used synthetic fungicides such as Agrosan G.N. and Captan (Pandey and Dubey 1992, 1994).

Pandey and Dubey (1994) determined the fungitoxic spectrum of *O. canum* oil (500 µl/l) and found 100% inhibition of the growth of the following fungi: *Fusarium oxysporum* f. sp. *ciceri*, *F. sesami*, *F. semitectum*, *Alternaria brassicae*, *A. solani*, *A. tenuissima*, *Cladosporium cladosporioides*, *Helminthosporium oryzae*, *Penicllium citrinum*, *Colletotrichum* sp. and *Drechslera auntii*.

O. sanctum leaf extract inhibited the radial growth of the rice pathogens *Pyricularia oryzae*, *Cochliobolus miyabeanus* (*Helminthosporium oryzae*) and *Rhizoctonia solani* (Tewari and Mandakini 1991). Asthana *et al.* (1982) found that the essential oil of *O. adscendens* was fungicidal and inhibited the growth of *Helminthosporium oryzae* at a concentration of 200 µl/l. The aqueous extract of *O. basilicum* leaves inhibited the growth of *Trichoconiella padwickii* in paddy (*Oryza sativa*) seeds (Shetty *et al.*, 1989).

A leaf extract of *O. canum* inhibited the germination of *Cercospora moricola*, a leaf spot causing pathogen on mulberry (Siddaramaiah and Hedge 1990).

A crude steam distillate of *O. gratissimum* sprayed onto infection courts on detached cocoa pods moments after inoculation with *Phytophthora palmivora* completely inhibited the pathogen and blackpod lesion development on 75% of the infection courts. In the field the extract also suppressed lesion development, although to a significantly lower extent and the fungitoxicity was lost within 3 h of application (Awuah 1994).

The essential oil of *O. adscendens* protected stored chilli (*Capsicum annuum*) seeds completely from fungal development and gave better control than the synthetic fungicides Bavistin, Blitox-50 and Dithane M-45 for 12 months (Asthana *et al.*, 1989).

Exudates of *O. basilicum* decreased the population of various fungi, including *Aspergillus* spp. and *Fusarium* spp. in the phyllosphere of beans (Afifi 1975). The essential oil of *O. basilicum* exhibited fungitoxic properties against aflatoxin-producing strains of *Aspergillus flavus* and *A. parasiticus*. The oil was fungistatic at a dose of 1.5 ml/l and fungicidal at 6.0 ml/l. These doses are much lower than those of commercial fungicides and fumigants, and the effect remained unaffected by temperature, storage, and increased inoculum. In addition, a dose of 1.5 ml/l was also effective against a number of other fungi listed in Table 5.2 with the superscript 26 (Dube *et al.*, 1989).

O. sanctum leaf extracts inhibited germination of *Pestalotia psidii* spores *in vitro* and on guava fruits dipped in the extract. The extract does not affect the fruit flavour (Pandey *et al.*, 1983). An aqueous leaf extract of *O. sanctum* gave good control of the disease development (*Botryodiplodia theobromae*, *Fusarium oxysporum*, *Helminthosporium spiciferum* [*Cochliobolus spicifer*], *Curvularia lunata* [*Cochliobolus lunatus*], *Aspergillus flavus* and *Trichothecium roseum*) in banana (Singh *et al.*, 1993). Sweet basil oil was highly effective in checking the growth of *Penicillium italicum* (blue mold) on tangerine (*Citrus reticulata*). The appearance and flavour of the fruits were unchanged (Arora and Pandey 1977).

Protection against fungi causing diseases in humans

The growth of the common fungus *Candida albicans* was inhibited by the essential oil of *O. gratissimum* showing an inhibition zone > 30 mm (Grover and Rao 1977, Ndounga and Ouamba 1997). The effect of *O. sanctum* on *C. albicans* was none according to Grover and Rao (1977) and according to Prasad *et al.* (1986) it inhibited the growth for 7 days. The essential oil of *O. basilicum* showed some inhibitory effect against *C. albicans* (Janssen *et al.*, 1986, Prasad *et al.*, 1986). However, the minimum inhibitory concentration was 1250 μg/ml compared to 312.5 μg/ml for *O. gratissimum* (Ndounga and Ouamba 1997).

A group of fungi often used to study the antifungal effects of different oils is the dermatophytes. Singh *et al.* (1985) report that the oil of *O. gratissimum* was active at 1000 ppm against *Epidermophyton floccosum*, *Microsporum canis* and *Trichophyton mentagrophytes*. Clinical isolates of *Trichophyton rubrum*, *T. mentagrophytes*, *M. canis* and *E. floccosum* were used to test the antifungal activity of the essential oil of *O. gratissimum*. It inhibited over 80% of the studied strains and produced inhibition zones > 10 mm in diameter (Lima *et al.*, 1992, 1993). A later report confirmed the effect of *O. gratissimum* against *T. rubrum* and *T. mentagrophytes* (Nwosu and Okafor 1995). Ndounga and Ouamba (1997) report a very good effect of *O. gratissimum* against *T. mentagrophytes*. The oil caused an inhibition zone of more than 30 mm. The minimum inhibitory concentration was 625 μg/ml. The oil of *O. gratissimum* was active against *T. mentagrophytes* var. *interdigitale* at a dilution of 1:6400 (Janssen *et al.*, 1989).

Essential oils of *O. basilicum* (French, Indian and Niazbo) were effective against *Trichophyton mentagrophytes*, *T. rubrum* and *T. verrucosum*, except for the Indian oil where some growth of *T. verrucosum* was detected within seven days (Prasad *et al.*, 1986). Janssen *et al.* (1988) determined maximum inhibitory dilutions for *O. basilicum* oil against *Epidermophyton floccosum*, *T. mentagrophytes* var. *interdigitale* and *T. rubrum* and report the values 1:3200–1:1600. On the other hand, Ndounga and Ouamba (1997) report the minimum inhibitory concentration of *O. basilicum* oil against *T. mentogrophytes* to be 5000 μg/ml.

The activity of other *Ocimum* species against *T. mentogrophytes* var. *interdigitale* is characterized by the maximum inhibitory dilution values as follows: *O. canum* 1:3200, *O. trichodon* 1:3200 and *O. urticifolium* 1:3200. One strain of *O. urticifolium* was active at the dilution 1:6400 (Janssen *et al.*, 1989). The essential oils of *O. sanctum* and O. *kilimandscharicum* were effective against *T. mentagrophytes*, *T. rubrum* and *T. verrucosum* without exceptions (Prasad *et al.*, 1986).

Individual oil components with antifungal properties are:

1. Eugenol, active against *Absidia glauca*, *Aspergillus nidulans*, *A. niger*, *Colletotrichum capsici*, *Fusarium moniliforme*, *Pestalotia psidii* and *Rhizopus nodosus* as a pure isolate and at dilutions 1:100 and 1:200 (Garg and Siddiqui 1992). At 0.1% concentration

eugenol oil (*O. gratissimum*) inhibited the growth of *Sclerotium rolfsii*, *Rhizoctonia* spp. and *Alternaria alternata* (Thakur *et al.*, 1989). *O. suave* oil (main component eugenol) inhibited the growth of *Saccharomyces cerevisae* with a minimum inhibitory concentration of 500 μg/ml (Chogo and Crank 1981).

2. Caryophyllene (pure isolate, 1:100 and 1:200) exhibited very strong antifungal activity against *Absidia glauca* (Garg and Siddiqui 1992).
3. 1,8-Cineole showed very good activity against *Alternaria alternata* and *Fusarium moniliforme* (Garg and Siddiqui 1992).
4. Thymol (*O. viride*) checked the growth of *Rhizoctonia solani* and *Sclerotium rolfsii* (Thakur *et al.*, 1989).
5. Linalool (*O. canum*) checked the growth of *Rhizoctonia solani* (Thakur *et al.*, 1989).

INSECTICIDAL ACTIVITY

In warm countries all kinds of insects are a nuisance as well as the cause of diseases in plants, animals and humans. Stored products has to be protected against different insects. Therefore much effort has been put into the development of different insect repellants. Since synthetic agents often have severe toxic effects and may be too expensive for people in developing countries much hope has been placed on insect repellants of plant origin, preferably plants growing locally. *Ocimum* species have been studied along with many others in this respect and the results of these studies are reported below.

The essential oil (2% in acetone) of *O. gratissimum* showed 100% repellency against the housefly, *Musca domestica* (Singh and Singh 1991).

The essential oil of *O. basilicum* showed repellent activity of class IV against red flour beetle, *Tribolium castaneum* (Mohiuddin *et al.*, 1987).

Protection of Growing Plants

The aphidicidal effect of an ethanol extract of *O. sanctum* was tested against the following aphid species: *Myzus persicae*, *Metopolophium dirhodum*, *Aphis fabae*, *Sitobion avenae* and *Acyrthosiphon pisum*. The best effect, with a mortality rate of 79%, was observed for *M. dirhodum* (Stein *et al.*, 1988).

Methanol extracts of leaves of *O. sanctum* at a concentration 1.0% resulted in 90% mortality of the grub of *Henosepilachna vigintioctopunctata*, a pest of brinjal, at 12 to 24h after treatment (Satpathi and Ghatak 1990).

Acetonic solutions with different concentrations of *O. basilicum* essential oil were sprayed on bean leaf discs and were found to induce repellency in adult females of the carmine spider mite, *Tetranychus cinnabarinus*. Also egglaying was reduced. The EC_{50}-value was 1.4% (Mansour *et al.*, 1986).

Fruit flies (*Dacus* spp.) cause heavy damage to ripe and semi-ripe fruits and vegetables. *Ocimum sanctum* extracts can be used to lure and trap fruit flies and so prevent damage to crops. Leaf extract in ethyl acetate (0.25 ml on a cotton pad) effectively attracted fruit flies from a distance of 0.8 km. Methyl eugenol gave the same result (Roomi *et al.*, 1993).

Protection of Stored Products

The dried ground leaves and essential oil of *O. kilimandscharicum* in doses of 25.0 g leaves and 0.3 g essential oil per 250 g grain (maize or sorghum) killed 100% of *Sitophilus zeamais*, *Rhyzopertha dominica* and *Sitotroga cerealella* in 48 h. The best repellent activity was seen by 0.3 g essential oil/250 g grains against *Sitophilus zeamais* (Jembere *et al.*, 1995).

Powdered leaves of *O. canum* at a concentration of 2% w/w completely inhibited the oviposition of adult *Zabrotes subfasciatus* in dried Pinto beans. 1% w/w powdered leaves in Pinto beans caused 100% mortality of adult *Z. subfasciatus* in 48 h. The EC_{50} was determined as 0.45% w/w. Linalool is the major compound in *O. canum* leaves and the fumigant toxicity can be explained by the gradual release of linalool from the powdered material. The contact toxicity evidently involves other components (Weaver *et al.*, 1994). Dose-response curves for linalool were completed with adult *Zabrotes subfasciatus*, *Acanthoscelides obtectus*, *Rhyzopertha dominica* and *Sitophilus oryzae* using a filter paper bioassay. The LC_{50}-values were: 428 $\mu g/cm^2$ (*Z. subfasciatus*), 405 $\mu g/cm^2$ (*A. obtectus*), 428 $\mu g/cm^2$ (*R. dominica*) and 427 $\mu g/cm^2$ (*S. oryzae*) (Weaver *et al.*, 1991). In another study the essential oil of *O. basilicum* was found to kill adult *Acanthoscelides obtectus* insects on kidney beans in the field and during storage, as well as inhibit reproduction through ovicidal and larvicidal effects (Regnault-Roger and Hamraoui 1994).

Powdered leaves of *O. gratissimum* reduced the egglaying of *Callosobruchus maculatus* in stored cowpea seeds (Ofuya 1990).

Protection of Animals

O. suave essential oil was found to repel and kill all stages of the tick *Rhipicephalus appendiculatus.* In an *in vitro* assay for the larvae the LC_{50} of the oil in liquid paraffin was 0.024%. A 10% solution was found to kill all immatures and more than 70% of the adults feeding on rabbits. The rabbits were protected for 5 days using the 10% solution. The same protection may be useful for field kept cattle (Mwangi *et al.*, 1995).

Protection of Humans Against Diseases

An extract of *O. sanctum* showed pupicidal effects against fleas, which are vectors of bubonic plaque, murine typhus and other rodent borne diseases (Renapurkar and Deshmukh 1984). A crude extract of *O. sanctum* showed pupicidal effect on newly emerged pupae of the vector *Aedes aegypti* (Kumari *et al.*, 1994). The essential oils of *O. basilicum* and *O. sanctum* and their major constituents showed insecticidal properties against the vectors *Anopheles stephensi*, *Aedes aegypti* and *Culex quinquefasciatus* in laboratory tests. *O. basilicum* and its major constituent, methyl chavicol, were more effective than *O. sanctum.* The essential oils and their major constituents were more toxic to *A. stephensi*, followed by *A. aegypti* and *C. quinquefasciatus* (Bhatnagar *et al.*, 1993).

OTHER ACTIVITIES

Plants belonging to the genus *Ocimum* exhibit a great deal of different pharmacological activities of which the most important, as concluded by the number of research reports, will be discussed below. The activities to be discussed in more detail are anti-inflammatory, immunomodulating and adaptogenic, anticarcinogenic, hypoglycemic and blood lipid lowering, radioprotective, effect on the CNS, antiulcerogenic, hepatoprotective and the effect on smooth muscle. In addition to these activities a number of other activities are also reported in the literature, such as antioxidant (Maulik *et al.*, 1997), angioprotective effect (Nikolaevskii *et al.*, 1990), effect on the reproductive behaviour (Kantak and Gogate 1992) and antiwormal activity (Jain and Jain 1972).

Anti-inflammatory Activity

Ocimum sanctum L., popularly known as "Tulsi" in Hindi and "Holy Basil" in English, is a widely known sacred plant of Hindus. Different parts of the plant have been claimed to be valuable in a wide spectrum of diseases (Singh *et al.*, 1996a). For instance, it is used for the treatment of arthritis, rheumatism, pain and fever in the Ayurvedic system of medicine (Godhwani *et al.*, 1987). *Ocimum sanctum* is now intensively studied in order to prove these activities by pharmacological evidence.

A methanol extract and an aqueous suspension of *O. sanctum* leaves inhibited acute as well as chronic inflammation in rats as tested by carrageenan-induced paw edema and granuloma pouch, respectively. In both test procedures, the anti-inflammatory response of methanol extract (500 mg/kg) was statistically equivalent to the response observed with 300 mg/kg of sodium salicylate. The methanol extract and the aqueous suspension showed a statistically significant antipyretic action. However, the antipyretic action was weaker and of shorter duration than that of 300 mg/kg of sodium salicylate. These effects may, at least in part, be attributed to inhibition of the biosynthesis of prostaglandins (Godhwani *et al.*, 1987).

O. sanctum leaves and essential oil exhibited analgesic activity when studied with the hot plate, tail flick, acetic acid writhing and naloxone antagonism methods. No allied activities, *e.g.* antipyretic or CNS-depressant, were detected (Vohora and Dandiya 1992).

A 50% aqueous ethanol extract of dried and fresh leaves, and the volatile and fixed oils of *O. sanctum* inhibited hind paw edema induced in rats by treatment with carrageenan, serotonin, histamine or PGE-2. The same extracts also showed anti-asthmatic activity against histamine- and acetylcholine-induced pre-convulsive dyspnea in guinea pigs (Singh and Agrawal 1991).

The anti-inflammatory activity of *O. sanctum* was studied using the water soluble portion of an alcoholic leaf extract. The test method used was carrageenan-induced paw edema in rats. The anti-inflammatory effect was dose dependent, showing a 57% inhibition at a dose of 400 mg/kg *i.p.* (Table 5.3). The same level of inhibition was found with 80 mg/kg phenylbutazone. The ED_{50}-value was 270 mg/kg and LD_{50} 4850 mg/kg, respectively (Chattopadhyay *et al.*, 1994).

Also the fixed oil of *O. sanctum* seeds has been screened for its anti-inflammatory potential. It possessed significant anti-inflammatory activity against carrageenan- and different other mediator-induced paw edemas in rats. On the basis of these findings it seems that *O. sanctum* fixed oil may have the potential to inhibit both the pathways causing inflammation, i.e. cyclooxygenase and lipoxygenase, of the arachidonic acid

Table 5.3 Anti-inflammatory activity of *Ocimum sanctum* leaf extract compared to phenylbutazone and saline control (Chattopadhyay *et al.*, 1994)

Treatment	*Dose* *mg/kg i.p.*	*Percent inhibition*
O. sanctum	50	11.11
	100	24.07
	200	40.74
	400	57.44
Phenylbutazone	20	29.62
	40	40.74
	60	53.70
	80	59.25
Control	2 ml/kg	–

metabolism (Singh *et al.*, 1996a). Later the same year it was found that the pharmacological activity of the fixed oil could be attributed to its triglyceride fraction or the fatty acids (Singh *et al.*, 1996b). A year later it was determined that linolenic acid present in the fixed oil of *O. sanctum* has the capacity to block both the cyclooxygenase and lipoxygenase pathways of the arachidonate metabolism (Singh and Majumdar 1997).

Immunomodulating and Adaptogenic Effects

Certain plants widely used in the Ayurvedic and Unani systems of medicine for treatment of chronic infections and immunological disorders were studied on delayed type hypersensitivity, humoral responses to sheep red blood cells, skin allograft rejection, and phagocytic activity of the reticuloendothelial system in mice. An ethanolic extract of *O. gratissimum* leaves (100 mg/kg *p.o.*) appeared to improve the phagocytic function without affecting the humoral or cell-mediated immune system (Atal *et al.*, 1986).

A methanol extract and an aqueous suspension of *O. sanctum* were investigated for their immunoregulatory profile to antigenic challenge in albino rats. The aqueous suspension represents the form usually employed in traditional medicine. Administration of the methanol extract at doses of 100 and 250 mg/kg (*p.o.*) and the aqueous suspension at 500 mg/kg caused a statistically significant increase in the antibody titre. This finding indicates that the methanol extract and the aqueous suspension of *O. sanctum* leaves stimulates humoral response and this observed immunostimulation may account for the adaptogenic action of the plant (Godhwani *et al.*, 1988).

The effects of alcoholic extracts of *Eleutherococcus senticosus* and *Ocimum sanctum*, regarded as anti-stress agents, were studied on the changes in central neurotransmitter (adrenaline, noradrenaline, dopamine, 5-hydroxytryptamine) levels and enzyme (monoamine oxidase) activity in the brain induced by stressors. There was an increase in the levels of dopamine and 5-hydroxytryptamine and decrease in adrenaline, noradrenaline and monoamine oxidase in stressed rats. Both plants prevented the decrease in adrenaline, noradrenaline and monoamine oxidase and facilitated the increase in dopamine. It was concluded that prevention of changes in brain neurotransmitters and enzymes by anti-stress agents appear to enhance the stress-adaption and the organism copes better with stressful situations (Singh *et al.*, 1991a). The anti-stress activity of *O. sanctum* was compared to *E. senticosus* and *Panax ginseng* in albino mice and rats in different experimental models of stress. All of them were found effective. An anti-stress unit was determined by the ED_{50}-values obtained in the tests. Taking the anti-stress unit of *O. sanctum* as 1, the relative effect of *E. senticosus* was 0.83 and *P. ginseng* 0.53. *O. sanctum* also had the highest margin of safety (Singh *et al.*, 1991b).

Effects of stress and its modulation by *O. sanctum* and eugenol were evaluated on some biochemical and biophysical parameters in rats. *O. sanctum* and eugenol lowered cholesterol levels and enzyme activities induced by stress. Also stress induced changes in membrane functions were affected by administration of *O. sanctum* and eugenol (Sen *et al.* 1992). The antistress effects of an ethanol extract of *O. sanctum* leaves was screened against acute and chronic noise stress in albino rats by investigating the plasma corticosterone level. Treatment of the animals with the *O. sanctum* extract prevented the changes in the plasma level of corticosterone (Sembulingan *et al.*, 1997).

Anticarcinogenic Effects

In a search for plant products against cancer the protective effect of ursolic acid from *O. sanctum* was studied against free radical induced damage. Ascorbic acid, carbon tetrachloride and ADP/iron were used to induce lipid peroxidation in isolated rat liver microsomes. Ursolic acid provided a 60% protection against lipid peroxidation (Balanehru and Nagarajan 1991). Ursolic acid was further studied against adriamycin induced lipid peroxidation in liver and heart microsomes *in vitro*. It showed 13% and 17% protection in liver and heart microsomes, respectively (Balanehru and Nagarajan 1992).

Basil leaves (*O. sanctum*) significantly decreased the incidence of both benzopyrene induced neoplasia and 3'-methyl-4-dimethylaminoazobenzene induced hepatomas in mice (Aruna and Sivaramakrishnan 1992).

An ethanolic extract of the leaves of *O. sanctum* showed chemopreventive effects on 7,12-dimethylbenzanthracene induced skin papillomagenesis in male Swiss albino mice. A significant reduction in the tumor incidence, average number of tumors/tumor bearing mice and the cumulative number of papillomas was observed in mice treated topically with the leaf extract of *O. sanctum* (Prashar *et al.*, 1994).

An alcoholic extract from the leaves of *O. sanctum* at doses of 400 and 800 mg/kg (*p.o.*) for 15 days significantly elevated the activity of cytochrome P-450, cytochrome b_5, aryl hydrocarbon hydroxylase and glutathione S-transferase in mice. The effect was dose-responsive. The above mentioned enzymes are important in the detoxification of carcinogens and mutagens. Glutathione S-transferase represents an important defense mechanism in protecting cells against oxygen-derived free radicals and also from cellular lethality after exposure to anticancer drugs or ionizing radiation. These findings suggest that *Ocimum* leaf extract has the potential to block or suppress the events associated with chemical carcinogenesis (Banerjee *et al.*, 1996).

Hypoglycemic and Blood Lipid Lowering Effect

One of the traditional uses for *O. sanctum* is for treating diabetes. Experimental studies performed in the 1960's showed that basil leaves can decrease blood glucose in hyperglycemic rats. These results were confirmed later by other studies which showed that oral administration of alcoholic extracts of *O. sanctum* leaves led to a marked lowering of blood sugar level in normal, glucose fed hyperglycemic and streptozotocin induced diabetic rats (Chattopadhyay 1993, Rai *et al.*, 1997). In a randomized placebo-controlled, single blind trial performed on humans a significant decrease in fasting and postprandial blood glucose levels was found. The medication consisted of 2.50 g dried leaf powder of fresh leaves of *O. sanctum* and was given for 61 days. The fasting blood glucose fell by 21.0 mg/dl and postprandial blood glucose fell by 15.8 mg/dl. The urine glucose levels showed a similar trend. In addition the total cholesterol showed a mild reduction during thc basil treatment period (Agrawal *et al.*, 1996).

Fresh leaves of *O. sanctum* (1 and 2 g/100 g diet) given for 4 weeks, brought about significant changes in the blood lipid profile of normal albino rats. Serum total cholesterol, triglycerides, phospholipids and LDL-cholesterol decreased and HDL-cholesterol and total faecal sterol content increased (Sarkar *et al.*, 1994).

Radioprotective Effect

Prompted by the adaptogenic properties of *O. sanctum* Devi and Ganasoundari (1995) undertook a study to investigate the radioprotective effect of *O. sanctum* leaf extracts. Fresh leaves were extracted with water or water-alcohol and the extracts were administered *i.p.* either as a single dose or multiple doses to albino mice before whole-body exposure to gamma radiation. The optimum dose of water extract was 50 mg/kg. The water extract was more effective and less toxic than the aqueous alcohol extract. The acute LD_{50}-values for the water and the aqueous alcohol extracts were 6200 mg/kg and 4600 mg/kg, respectively.

O. sanctum aqueous extract seems to provide protection against changes in the bone marrow and chromosome damage induced by radiation. Albino mice receiving both *O. sanctum* extract and radiation showed fewer changes than groups receiving only radiation. The mechanism for protection is likely based on free radical scavenging (Ganasoundari *et al.*, 1997a, 1997b).

Effects on the Central Nervous System

An ethanolic extract of the leaves of *O. sanctum* was screened for its effects on the central nervous system. It prolonged the sleeping time induced by pentobarbital, reduced the recovery time and severity of elektroshock- and pentylenetetrazole-induced convulsions and decreased the apomorphine-induced fighting response and open-field activity. All these actions resemble the activity of low doses of barbiturates. The mechanism of action for the *Ocimum* extract may involve dopaminergic neurones (Sakina *et al.*, 1990).

Antiulcerogenic Effect

Aqueous and methanolic extracts of *O. basilicum* showed antiulcerogenic effects when administered to rats with aspirin-induced gastric ulcers. The ulcer index was decreased by both extracts. The acid output was decreased only by the methanolic extract, which indicate that the active principles are soluble in methanol. The pepsin output was decreased by both the aqueous and methanolic extract in ulcerated, but not in normal animals. The aqueous extract of *O. basilicum* also increased the hexosamine concentration in normal rats, which might contribute to its antiulcer effect (Akhtar and Munir 1989).

The antiulcerogenic effects of extracts, volatile oils and flavonoid glycosides of *O. basilicum* leaves were studied in normal as well as aspirin-, acetic acid- and stress-induced ulcerated rats. Their effect on the output of gastric acid, pepsin and hexosamines was recorded. The aqueous, methanol and water-methanol extracts and flavonoid glycosides decreased the ulcer index, inhibited gastric acid and pepsin secretion and enhanced hexosamines. It appears that the antiulcerogenic compounds of *O. basilicum* are extractable both into water and methanol and that they may include flavonoid glycosides. These substances may act by augmenting the gastric barrier (Akhtar *et al.*, 1992).

A study on *O. sanctum* showed that the extract reduced the ulcer index, free and total acidity on acute and chronic administration. It also increased the mucous secretion (Mandal *et al.* 1993).

Hepatoprotective Activity

O. sanctum leaves are part of a preparation sold in India for liver ailments (Handa *et al.*, 1986). To prove the hepatoprotective action *O. sanctum* leaf extract (200 mg/kg orally) was given to rats with paracetamol induced hepatic damage. The serume enzyme levels were significantly reduced and liver GSH level significantly higher in animals receiving both paracetamol and *O. sanctum* leaf extract than in those given paracetamol alone. Histopathological studies showed marked reduction in fatty degeneration in animals receiving both paracetamol and *O. sanctum* compared to the control group (Chattopadhyay *et al.*, 1992).

O. gratissimum and *O. basilicum* are used in a Taiwanese herbal remedy believed to possess anti-inflammatory and detoxication activities. The crude extracts was studied against CCl_4- and D-GalN-induced acute hepatitis and were found to be hepatoprotective (Lin *et al.*, 1995).

Effects on Smooth Muscle

A crude lipid extract of *O. gratissimum* leaves caused contractions in a guinea pig ileum, rat colon and raised the mean arterial blood pressure in rats. The compounds seem to be related to lipids and fairly polar (Onajobi 1986).

The volatile oil of basil (*O. basilicum*) had relaxant effects on tracheal and ileal smooth muscles of the guinea pig. Of 22 oils examined basil belonged to the most potent group consisting of 16 oils. The EC_{50}-value for tracheal muscle was 19 mg/l and for ileal muscle 32 mg/l, respectively (Reiter and Brandt 1985).

REFERENCES

Abdel-Sattar, A., Bankova, V., Kujumgiev, A., Galabov, A., Ignatova, A., Todorova, C. and Popov, S. (1995) Chemical composition and biological activity of leaf exudates from some Lamiaceae plants. *Pharmazie*, **50**, 62–65.

Afifi, A.F. (1975) Effect of volatile substances from species of Labiatae on rhizospheric and phyllospheric fungi of *Phaseolus vulgaris. Phytopatologische Zeitschrift*, **83**, 296–302.

Agrawal, P., Rai, V. and Singh, R.B. (1996) Randomized placebo-controlled, single blind trial of holy basil leaves in patients with noninsulin-dependent *diabetes mellitus. Int. J. Clin. Pharm. Ther.*, **34**, 406–409.

Akhtar, M.S.A., Akhtar, A.H. and Khan, M.A. (1992) Antiulcerogenic effects of *Ocimum basilicum* extracts, volatile oils and flavonoid glycosides in albino rats. *Int. J. Pharmacognosy*, **30**, 97–104.

Akhtar, M.S. and Munir, M. (1989) Evaluation of the gastric antiulcerogenic effects of *Solanum nigrum, Brassica oleraceae* and *Ocimum basilicum* in rats. *J. Ethnopharmacol.*, **27**, 163–176.

Arora, R. and Pandey, G.N. (1977) The application of essential oils and their isolates for blue mold decay control in *Citrus reticulata* Blanco. *J. Food Sci. Technol.*, **14**, 14–16.

Aruna, K. and Sivaramakrishnan, V.M. (1992) Anticarcinogenic effects of some Indian plant products. *Food Chem. Toxicol.*, **30**, 953–956.

Asthana, A., Chandra, H., Dikshit, A. and Dixit, S.N. (1982) Volatile fungitoxicants from leaves of some higher plants against *Helminthosporium oryzae. Zeitschrift für Pflanzenkrankheiten und Pflanzenschutz*, **89**, 475–479.

Asthana, A., Dixit, K., Tripathi, N.N. and Dixit, S.N. (1989) Efficacy of *Ocimum* oil against fungi attacking chilli seed during storage. *Tropical Science*, **29**, 15–20.

Asthana, A., Tripathi, N.N. and Dixit, S.N. (1986) Fungitoxic and phytotoxic studies with essential oil of *Ocimum adscendens. Journal of Phytopathology*, **117**, 152–159.

Atal, C.K., Sharma, M.L., Kaul, A. and Khajuria, A. (1986) Immunomodulating agents of plant origin. I. Preliminary screening. *J. Ethnopharmacol.*, **18**, 133–141.

Awasthi, S.K. (1991) Response of toxin-producer and non-toxin producer strains of *Aspergillus flavus* (Link.) towards various volatile constituents. *Geobios*, **10**, 84–85.

Awuah, R.T. (1994) In vivo use of extracts from *Ocimum gratissimum* and *Cymbopogon citratus* against *Phytophtora palmivora* causing blackpod disease of cocoa. *Annals Appl. Biol.*, **124**, 173–178.

Balanehru, S. and Nagarajan, B. (1991) Protective effect of oleanolic acid and ursolic acid against lipid peroxidation. *Biochemistry International*, **24**, 981–990.

Balanehru, S. and Nagarajan, B. (1992) Intervention of adriamycin induced free radical damage. *Biochemistry International*, **28**, 735–744.

Banerjee, S., Prashar, R., Kumar, A. and Rao, A.R. (1996) Modulatory influence of alcoholic extract of *Ocimum* leaves on carcinogen-metabolizing enzyme activities and reduced glutathione levels in mouse. *Nutr. Cancer*, **25**, 205–217.

Bhatnagar, M., Kapur, K.K., Jalees, S. and Sharma, S.K. (1993) Laboratory evaluation of insecticidal properties of *Ocimum basilicum* Linnaeus and *O. sanctum* Linnaeus plants essential oils and their major constituents against vector mosquito species. *J. Entomol. Res.*, **17**, 21–26.

Caceres, A, Cano, O., Samayoa, B. and Aguilar, L. (1990) Plants used in Guatemala for the treatment of gastrointestinal disorders. 1. Screening of 84 plants against enterobacteria. *J. Ethnopharmacol.*, **30**, 55–73.

Chattopahdyay, R.R. (1993) Hypoglycemic effect of *Ocimum sanctum* leaf extract in normal and streptozotocin diabetic rats. *Indian J. Exp. Biol.*, **31**, 891–893.

Chattopahdyay, R.R., Sarkar, S.K., Ganguly, S. and Basu, T.K. (1994) A comparative evaluation of some anti-inflammatory agents of plant origin. *Fitoterapia*, **LXV**, 146–148.

Chattopahdyay, R.R., Sarkar, S.K., Ganguly, S., Medda, C. and Basu, T.K. (1992) Hepatoprotective activity of *Ocimum sanctum* leaf extract against paracetamol induced hepatic damage in rats. *Indian J. Pharmacol.*, **24**, 163–165.

Chogo, J.B. and Crank, G. (1981) Chemical composition and biological activity of the Tanzanian plant *Ocimum suave*. *Lloydia*, **44**, 308–311.

Devi, P.U. and Ganasoundari, A. (1995) Radiprotective effect of leaf extract of Indian medicinal plant *Ocimum sanctum*. *Ind. J. Exp. Biol.*, **33**, 205–208.

Dey, B.B. and Choudhuri, M.A. (1984) Essential oil of *Ocimum sanctum* L. and its antimicrobial activity. *Indian Perfumer*, **28**, 82–87.

Dube, S., Upadhyay, P.D. and Tripathi, S.C. (1989) Antifungal, physicochemical, and insect-repelling activity of the essential oil of *Ocimum basilicum*. *Can. J. Bot.*, **67**, 2085–2087.

El-said, F., Sofowora, E.A., Malcolm, S.A. and Hofer, A. (1969) An investigation onto the efficacy of *Ocimum gratissimum* as used in Nigerian native medicine. *Planta Medica*, **17**, 195–200.

Ganasoundari, A., Devi, P.U. and Rao, M.N.A. (1997a) Protection against radiation-induced chromosome damage in mouse bone marrow by *Ocimum sanctum*. *Mut. Res.*, **373**, 271–276.

Ganasoundari, A., Zare, S.M. and Devi, P.U. (1997b) Modification of bone marrow radiosensitivity by medicinal plant extracts. *British J. of Radiology*, **70**, 599–602.

Garg, S.C. and Siddiqui, N. (1992) Antifungal activity of some essential oil isolates. *Pharmazie*, **47**, 467–468.

Girón, L.M., Freire, V., Alonzo, A. and Cáceres, A. (1991) Ethnobotanical survey of the medicinal flora used by the Caribs of Guatemala. *J. Ethnopharmacol.*, **34**, 173–187.

Githinji, C.W. and Kokwaro, J.O. (1993) Ethnomedicinal study of major species in the family Labiatae from Kenya. *J. Ethnopharmacol.*, **39**, 197–203.

Godhwani, S., Godhwani, J.L. and Vyas, D.S. (1987) *Ocimum sanctum*: an experimental study evaluating its anti-inflammatory, analgesic and antipyretic activity in animals. *J. Ethnopharmacol.*, **21**, 153–163.

Godhwani, S., Godhwani, J.L. and Vyas, D.S. (1988) *Ocimum sanctum* - a preliminary study evaluating its immunoregulatory profile in albino rats. *J. Ethnopharmacol.*, **24**, 193–198.

Grover, G.S. and Rao, J. T. (1977) Untersuchungen über die antimikrobielle Wirksamkeit der ätherischen öle von *Ocimum sanctum* und *Ocimum gratissimum*. *Parfümerie und Kosmetik*, **58**, 326–328.

Handa, S.S., Sharma, A. and Chakraborty, K.K. (1986) Natural products and plants as liver protecting drugs. *Fitoterapia*, **LVII**, 307–345.

Ilori, M.O., Sheteolu, A.O., Omonigbehin, E.A. and Adeneye, A.A. (1996) Antidiarrhoeal activities of *Ocimum gratissimum* (Lamiaceae). *J. Diarrhoeal Dis. Res.*, **14**, 283–285.

Jain, M.L. and Jain, S.R. (1972) Therapeutic utility of *Ocimum basilicum* var. *album*. *Planta Medica*, **22**, 66–70.

Janssen, A.M., Chin, N.L.J., Scheffer, J.J.C. and Baerheim Svendsen, A. (1986) Screening for antimicrobial activity of some essential oils by the agar overlay technique. *Pharmaceutisch Weekblad, Scientific edition*, **8**, 277–280.

Janssen, A.M., Scheffer, J.J.C., Parhan-van Atten, A.W. and Baerheim Svendsen, A. (1988) Screening of some essential oils for their activities on dermatophytes. *Pharmaceutisch Weekblad, Scientific edition*, **10**, 289–292.

Janssen, A.M., Scheffer, J.J.C., Ntezurubanza, L. and Baerheim Svendsen, A. (1989) Antimicrobial activities of some *Ocimum* species grown in Rwanda. *J. Ethnopharmacol.*, **26**, 57–63.

Jedlickova, Z., Mottl, O. and Sery, V. (1992) Antibacterial properties of the Vietnamese cajeput oil and *Ocimum* oil in combination with antibacterial agents. *J. Hyg. Epidemiol. Microbiol. Immunol.*, **36**, 303–309.

Jembere, B., Obeng-Ofori, D., Hassanali, A. and Nyamasyo, G.N.N. (1995) Products derived from the leaves of *Ocimum kilimandscharicum* (Labiatae) as post-harvest grain protectants against the infestation of three major stored product insect pests. *Bull. Entomol. Res.*, **85**, 361–367.

Kantak, N.M. and Gogate, M.G. (1992) Effect of short term administration of Tulsi (*Ocimum sanctum* Linn.) on reproductive behaviour of adult rats. *Ind. J. Physiol. Pharmacol.*, **36**, 109–111.

Kumari, C.P., Sharma, C.L. and Saxena, R.C. (1994) Pupicidal effect of *Ocimum sanctum* on the vector *Aedes aegypti* (Diptera: Culicidae). *J. Ecobiol.*, **6**, 69–70.

Lakshmanan, K.K. and Sankara Narayanan, A.S. (1990) Antifertility herbals used by the tribals in Anaikkatty Hills, Coimbatore District, Tamil Nadu. *J. Econ. Tax. Bot.*, **14**, 171–173.

Leung, A.Y. and Foster, S. (1996) *Encyclopedia of Common Natural Ingredients Used in Food, Drugs, and Cosmetics*, 2nd edition, John Wiley & Sons, New York, USA.

Lima, E.O., Gompertz, O.F., Paulo, M.Q. and Giesbrecht, A.M. (1992) *In vitro* antifungal activity of essential oils against clinical isolates of dermatophytes. *Rev. Microbiol.*, **23**, 235–248.

Lima, E.O., Gompertz, O.F., Giesbrecht, A.M. and Paulo, M.Q. (1993) *In vitro* antifungal activity of essential oils obtained from officinal plants against dermatophytes. *Mycoses*, **36**, 333–336.

Lin, C.C., Lin, J.K. and Chang, C.H. (1995) Evaluation of hepatoprotective effects of "Chhit-Chan-Than" from Taiwan. *Int. J. Pharmacognosy*, **33**, 139–143.

Mandal, S., Das, D.N., De, K., Ray, K., Roy, G., Chaudhuri, S.B., Sahana, C.C. and Chowdhuri, M.K. (1993) *Ocimum sanctum* Linn. A study on gastric ulceration and gastric secretion in rats. *Indian J. Physiol. Pharmacol.*, **37**, 91–92.

Mansour, F., Ravid, U. and Putievsky, E. (1986) Studies of the effect of essential oils isolated from 14 species of Labiatae on the carmine spider mite, *Tetranychus cinnabarinus. Phytoparasitica*, **14**, 137–142.

Maulik, G., Maulik, N., Bhandari, V., Kagan, V.E., Pakrashi, S. and Das, D.K. (1997) Evaluation of antioxidant effectiveness of a few herbal plants. *Free Radical Research*, **27**, 221–228.

Mishra, A.K., Dwivedi, S.K. and Kishore, N. (1989) Antifungal activity of some essential oils. *National Academy Science Letters*, **12**, 335–336.

Mohiuddin, S., Qureshi, R.A., Khan, M.A. and Nasir, M.K.A. (1987) Laboratory investigations on the repellency of some plant oils to red flour beetle, *Tribolium castaneum* Herbst. *Pak. J. Sci. Ind. Res.*, **30**, 754–756.

Monograph, Basilici herba. *Bundesanzeiger*, March **18**, 1992.

Mwangi, E.N., Hassanali, A., Essuman, S., Myandat, E., Moreka, L. and Kimondo, M. (1995) Repellant and acaricidal properties of *Ocimum suave* against *Rhipicephalus appendiculatus* ticks. *Exp. Appl. Acarology*, **19**, 11–18.

Ndounga, M. and Ouamba, J.M. (1997) Antibacterial and antifungal activities of essential oils of *Ocimum gratissimum* and *O. basilicum* from Congo. *Fitoterapia*, **LXVIII**, 190–191.

Nikolaevskii, V.V., Kononova, N.S., Pertsovskii, A.I. and Shinkarchuk, I.F. (1990) Effect of essential oils on the course of experimental atherosclerosis. *Patol. Fiziol. Eksp. Ter.*, **5**, 52–53.

Nwosu, M.O. and Okafor, J.I. (1995) Preliminary studies of the antifungal activities of some medicinal plants against *Basidiobolus* and some other pathogenic fungi. *Mycoses*, **38**, 194–195.

Ofuya, T.I. (1990) Oviposition deterrence and ovicidal properties of some plant powders against *Callosobruchus maculatus* in stored cowpea (*Vigna unguiculata*) seeds. *J. Agric. Sci.*, **115**, 343–346.

Onajobi, F.D. (1986) Smooth muscle contracting lipid-soluble principles in chromatographic fractions of *Ocimum gratissimum. J. Ethnopharmacol.*, **18**, 3–11.

Pandey, R.S., Bhargava, S.N., Shukla, D.N. and Dwided, D.K. (1983) Control of Pestalotia fruit rot of guava by leaf extracts of two medicinal plants. *Rev. Mex. Phytopatol.*, **2**, 15–16.

Pandey, V.N. and Dubey, N.K. (1992) Effect of essential oils from some higher plants against fungi causing damping-off disease. *Biologia Plantarum*, **34**, 143–147.

Pandey, V.N. and Dubey, N.K. (1994) Antifungal potential of leaves and essential oils from higher plants against soil phytopathogens. *Soil Biol. Biochem.*, **26**, 1417–1421.

Prakash, R. and Singh, V.A. (1986) Fungicidal activity of essential plant oils. *Agric. Biol. Res.*, **2**, 41–43.

Prasad, G., Kumar, A., Singh, A.K., Bhattacharya, A.K., Singh, K. and Sharma, V.D. (1986) Antimicrobial activity of essential oils of some *Ocimum* species and clove oil. *Fitoterapia*, **LVII**, 429–432.

Prashar, R., Kumar, A., Banerjee, S. and Rao, A.R. (1994) Chemopreventive action by an extract from *Ocimum sanctum* on mouse skin papillomagenesis and its enhancement of skin glutathione S-transferase activity and acid soluble sulfhydryl level. *Anti-cancer Drugs*, **5**, 567–572.

Rai, V., Iyer, U. and Mani, U.V. (1997) Effect of Tulasi (*Ocimum sanctum*) leaf powder supplementation on blood sugar levels, serum lipids and tissue lipids in diabetic rats. *Plant Foods for Human Nutrition*, **50**, 9–16.

Ramanoelina, A.R., Terrom, G.P., Bianchini, J.P. and Coulanges, P. (1987) Antibacterial action of essential oils extracted from Madagascar plants. *Arch. Inst. Pasteur Madagascar*, **53**, 217–226.

Regnault-Roger, C. and Hamraoui, A. (1994) Inhibition of reproduction of *Acanthoscelides obtectus* Say (Coleoptera), a kidney bean (*Phaseolus vulgaris*) bruchid, by aromatic essential oils. *Crop Protection*, **13**, 624–628.

Reiter, M. and Brandt, W. (1985) Relaxant effects on tracheal and ileal smooth muscles of the guinea pig. *Arzneimittelforschung*, **35**, 408–414.

Renapurkar, D.M. and Deshmukh, P.B. (1984) Pulicidal activity of some indigenous plants. *Insect. Sci. Appl.*, **5**, 101–102.

Roomi, M.W., Abbas, T., Shah, A.H., Robina, S., Qureshi, S.A., Hussain, S.S. and Nasir, K.A. (1993) Control of fruit-flies (*Dacus* spp.) by attractants of plant origin. *Anzeiger für Schädlingskunde Fflanzenschutz Umweltschutz*, **66**, 155–157.

Sakina, M.R., Dandiya, P.C., Hamdard, M.E. and Hameed, A. (1990) Preliminary psychopharmacological evaluation of *Ocimum sanctum* leaf extract. *J. Ethnopharmacol.*, **28**, 143–150.

Saksena, N. and Tripathi, H.H.S. (1985) Plant volatiles in relation to fungistasis. *Fitoterapia*, **LVI**, 243–244.

Sarkar, A., Lavania, S.C., Pandey, D.N. and Pant, M.C. (1994) Changes in the blood lipid profile after administration of *Ocimum sanctum* (Tulsi) leaves in the normal albino rabbits. *Indian J. Physiol. Pharmacol.*, **38**, 311–312.

Satpathi, C.R. and Ghatak, S.S. (1990) Evaluation of the efficacy of some indigenous plant extracts against *Henosepilachna vigintioctopunctata* (Coccinellidae: Coleoptera), a pest of brinjal. *Environ. Ecol.*, **8**, 1287–1289.

Sembulingan, K., Sembulingan, P. and Namasivayam, A. (1997) Effect of *Ocimum sanctum* Linn. on noise induced changes in plasma corticosterone level. *Indian J. Physiol. Pharmacol.*, **41**, 139–143.

Sen, P., Maiti, P.C., Puri, S., Ray, A., Audulov, N.A. and Valdman, A.V. (1992) Mechanism of anti-stress activity of *Ocimum sanctum* Linn., eugenol and *Tinospora malabrica* in experimental animals. *Indian J. Exp. Biol.*, **30**, 592–596.

Shetty, S.A., Prakash, H.S. and Shetty, H.S. (1989) Efficacy of certain plant extracts against seed-borne infection of *Trichoconiella padwickii* in paddy (*Oryza sativa*). *Can. J. Bot.*, **67**, 1956–1958.

Siddaramaiah, A.L. and Hedge, R.K. (1990) Effect of plant extract on *Cercospora moricola*: A leaf spot causing pathogen on mulberry. *Mysore Journal of Agricultural Sciences*, **24**, 208–213.

Singh, S. and Agrawal, S.S. (1991) Anti-asthmatic and anti-inflammatory activity of *Ocimum sanctum*. *Int. J. Pharmacognosy*, **29**, 306–310.

Singh, S. and Majumdar, D.K. (1997) Evaluation of antiinflammatory activity of fatty acids of *Ocimum sanctum* fixed oil. *Indian J. Exp. Biol.*, **35**, 380–383.

Singh, S., Majumdar, D.K. and Rehan, H.M.S. (1996) Evaluation of anti-inflammatory potential of fixed oil of *Ocimum sanctum* (Holybasil) and its possible mechanism of action. *J. Ethnopharmacol.*, **54**, 19–26.

Singh, S., Majumdar, D.K. and Yadav, M.R. (1996) Chemical and pharmacological studies on fixed oil of *Ocimum sanctum*. *Indian J. Exp. Biol.*, **34**, 1212–1215.

Singh, N., Misra, N., Srivastava, A.K., Dixit, K.S. and Gupta, G.P. (1991a) Effect of anti-stress plants on biochemical changes during stress reaction. *Indian J. Pharmacol.*, **23**, 137–142.

Singh, H.N.P., Prasad, M.M. and Sinha, K.K. (1993) Efficacy of leaf extracts of some medicinal plants against disease development in banana. *Letters in Applied Microbiology*, **17**, 269–271.

Singh, D. and Singh, A.K. (1991) Repellent and insecticidal properties of essential oils against housefly, *Musca domestica*. *Insect. Sci. Appl.*, **12**, 487–491.

Singh, S.P., Singh, S.K. and Tripathi, S.C. (1985) Antifungal activity of essential oils of some Labiatae plants against dermatophytes. *Indian Perfumer*, **27**, 171–173.

Singh, N., Verma, P., Mischra, N. and Nath, R. (1991b) A comparative evaluation of some anti-stress agents of plant origin. *Ind. J. Pharmacol.*, **23**, 99–103.

Sinha, P. and Saxena, S.K. (1989) Effect of treating tomatoes with leaf extracts of certain plants on the development of fruit rot caused by *Aspergillus niger* in the presence of *Drosophila busckii*. *Journal of Phytological Research*, **2**, 97–102.

Stein, U., Sayampol, B., Klingauf, F. and Bestmann, H.J. (1988) Aphidicidal effect of ethanol extracts of the holy basil, *Ocimum sanctum*. *Entomol. Gen.*, **13**, 229–237.

Tewari, S.N. and Mandakini, N. (1991) Activity of four plant leaf extracts against three fungal pathogens of rice. *Tropical Agriculture*, **68**, 373–375.

Thakur, R.N., Singh, P. and Khosla, M.K. (1989) *In vitro* studies on antifungal activities of some aromatic oils. *Indian Perfumer*, **33**, 257–260.

Thomas, O.O. (1989) Re-examination of the antimicrobial activities of *Xylopia aethiopica*, *Carica papaya*, *Ocimum gratissimum* and *Jatropha curcas*. *Fitoterapia*, **LX**, 147–155.

Tripathi, R.D., Banerji, R., Sharma, M.L., Balasub-Rahmanyam, V.R. and Nigam, S.K. (1985) Toxicity of essential oil from a new strain of *Ocimum gratissimum* (Cl*ocimum*) against betelvine pathogenic fungi. *Agric. Biol. Chem.*, **49**, 2277–2282.

Upadhyaya, M.L. and Gupta, R.C. (1990) Effect of extracts of some medicinal plants on the growth of *Curvularia lunata*. *Indian J. Mycol. Plant Pathol.*, **20**, 144–145.

Weaver, D.K., Dunkel, F.V., Ntezurubanza, L. and Jackson, L.L. (1991) The efficacy of linalool of freshly-milled *Ocimum canum* Sims. (Lamiaceae), for protection against post-harvest damage by certain stored product Coleoptera. *J. Stored Prod. Res.*, **27**, 213–220.

Weaver, D.K., Dunkel, F.V., Potter, R.C. and Ntezurubanza, L. (1994) Contact and fumigant efficacy of powdered and intact *Ocimum canum* Sims. (Lamiales:Lamiaceae) against Zabrotes subfasciatus (Boheman) adults (Coleoptera: Bruchidae). *J. Stored. Prod. Res.*, **30**, 243–252.

Wichtl, M. (1989) *Teedrogen*, 2. Auflage, Wissenschaftliche Verlagsgesellschaft mbH, Stuttgart, Germany.

Vohora, S.B. and Dandiya, P.C. (1992) Herbal analgesic drugs. *Fitoterapia*, **LXIII**, 195–207.

6. PROCESSING AND USE OF BASIL IN FOODSTUFFS, BEVERAGES AND IN FOOD PREPARATION

SEIJA MARJATTA MÄKINEN[1] and KIRSTI KAARINA PÄÄKKÖNEN[2]

[1]*Department of Applied Chemistry and Microbiology, Division of Nutrition, P.O Box 27, FIN-00014 University of Helsinki, Finland*
[2]*Department of Food Technology, P.O Box 27, FIN-00014 University of Helsinki, Finland*

Use of herbs in food preparation is as old as food preparation itself. Already as a forager, man was collecting aromatic, good-tasting herbs for his food and drink. Generally they were used to give more aroma, but sometimes also to cover the deteriorated taste and smell of the foodstuff. Man started to grow herbs during the transition from foraging to agriculture.

Herbs originate mainly from Mediterranean countries and from South Asia. Long before tea arrived in Europe, hot drinks were prepared from herbs. Having its origins in India and Iran, sweet basil spread into the Mediterranean countries, where it still grows wild. The name basil comes from the Greek *basilikos*, meaning royal. Only the sovereign (*basileus*) was allowed to cut basil. Basil belongs to the traditional wedding ceremony of the Hindu religion. According to African fairy tales, no poison could affect a man who had eaten basil. It has been a holy plant in India bringing good luck, a symbol of sorrow and hatred in Greece, and a herb for lovers in Italy. A basil pot at a girl's window sill meant an invitation to her lover to visit. Basil is still one of the most common herbs in the Italian kitchen. In southern Russia tradition has it that if a woman gives a sprig of basil to a man, and the man takes it willingly, he will love that woman for ever. Basil has indeed been an ingredient of love drinks and of aphrodisiac dishes. Ancient Romans used to cast leaves of basil on the floor of their dining room in order to excite their appetite with the odor of basil. As early as in 1775 the early American *Virginia Gazette* was advertizing basil.

In Finland sweet basil has been cultivated since the seventeenth century, first in monasteries and in big manor houses and later in the gardens of bigger farm houses. However, the beginning of the 1900's brought great famine years and wars, and herb gardens disappeared from Finland. The last decade has brought herbs back into home gardens, and herb gardens are now popular in farmhouses as well as in summer cottage gardens. Nowadays in Finland there are ca. 45 greenhouse growers who grow herbs for retail stores to sell in pots. More than 3 million pots of dill, parsley and chives are sold annually in Finland, and sweet basil is fourth in popularity; 34% of the remaining 10 herbs sold in pots are basil.

The consumption of herbs has increased in the 1990's. Health conscious consumers try to decrease their fat intake, and one of the means to get more aroma in food is to use herbs and spices instead of fat. In Europe the overall demand for dry culinary herbs is greatest in France, with 75% being imported (Maftei 1992). In the USA the most popular herbs are oregano, basil and sage (Friedman 1993). The increase of tourism has brought new popular dishes like pizzas and pasta dishes into the western food culture. The herb spices that are necessary in them are one of the reasons for their popularity and for their increased use in catering and in the food industry.

IS BASIL A NUTRITIONALLY VALUABLE HERB?

We do not eat herbs in such amounts that they would be a significant source of vitamins or minerals in our daily diet. According to McCance and Widdowson's food composition tables (Holland *et al.*, 1991) 100 grams of fresh basil leaves contain 26 mg of vitamin C, 3.95 mg of carotene, 0.08 mg of thiamin, 0.31 mg of riboflavin, and 1.1 mg of niacin. Its potassium content is 30 times that of sodium. One hundred grams of fresh basil leaves contain 250 mg of calcium, 37 mg of phosphorus, 5.5 mg of iron, and 11 mg of magnesium. According to Japanese investigations (Yamawaki *et al.*, 1993), the vitamin C content of basil is 67 ± 19 mg/100 g fresh leaves, of which nearly half is lost when heated for 5 min in boiling water. According to Chen *et al.* (1993) the β-carotene content of sweet basil is nearly twice that of carrot. The seeds of *Ocimum basilicum* are high in fiber and they can be considered a new source of dietary fiber. Indian scientists (Mathews *et al.*, 1993) have made a preparation called "**Falooda**", which contains basil seeds with water or milk made as a thickened mucilaginous food.

BASIL HAS EFFECTIVE ANTIOXIDATIVE PROPERTIES

Recent research has shown that sweet basil is an effective antioxidant. In studies done by Felice *et al.* (1993) basil showed higher antioxidant effectiveness than mint, oregano, parsley, rosemary, sage, chili, onions or garlic, when added to sunflower oil at a level 5 g/250 g oil. Its antioxidant effectiveness was even more than that of BHA (butylhydroxyanisol) or BHT (butylhydroxytoluene). There are many substances responsible for the antioxidant effect of basil, such as β-carotene, (Chen *et al.*, 1993); tocopherol, the content of which in basil according to Saito and Asari (1976) is greater than 100 mg %; and eugenol, isoeugenol (Klaui 1973), linalool and linalyl acetate (Yan and White 1990, Beckstrom-Sternberg and Duke 1994) as well as flavonoids. Eugenol, however, has a strong offensive smell, and its use as an antioxidant in foods is therefore limited. Isoeugenol, which occurs also as a natural substance in basil, has a more pleasant smell and is more preferable (Klaui 1973). According to a Japanese study (Saito *et al.*, 1976) petroleum ether insoluble fraction of basil has an antioxidant effect equal to that of 0.1% tocopherol. In Egypt Bassiouny *et al.* (1990) suggest that purified ether extracts of basil (0.02% w/w in dough) could be used as replacements for conventional antioxidants in soda cracker biscuits; the extracts did not have any effect on organoleptic properties. Powdered leaves had an antioxidant effect as well. Russian scientists (Fomicheva *et al.*, 1982) have shown that basil added to food concentrates or to chocolate inhibited peroxide formation and was able to decompose peroxides already formed.

BASIL IS A POTENTIAL ANTICARCINOGENIC AGENT IN OUR DIET

Although basil does not play any important role in our daily diet as a source of vitamins or minerals, it may have other significant effects in our nutrition. Aruna and Sivaramakrishnan (1992) have studied the anticarcinogenic properties of basil leaves (*Ocimum sanctum* L.) on mice and rats with carcinoma. Basil leaves and cumin seeds significantly decreased the incidence of BaP- (benzo[a]pyrene) induced neoplasia and 3'MeDAB (3'-methyl-4-dimethylaminoazobenzene)-induced hepatomas. They suggested that basil leaves, which are widely used in Indian cooking, may prove to be valuable anticarcinogenic agents. Genotoxicity (Tateo *et al.*, 1989) and mutagenity (Bersani *et al.*, 1981) of basil have been investigated in Italy with *Saccharomyces cerevisiae* and *Salmonella typhimurium*, respectively, with no positive findings. In the book **Spices, Herbs and Edible Fungi** Charalambous (1994) gives a list of pharmacologically and physiologically effective substances in herbs, basil included.

BASIL CONTAINS ANTIMICROBIAL SUBSTANCES

In addition to antioxidative, anticarcinogenic and antimutagenic substances, basil contains also antimicrobial compounds, one of them being eugenol (Meena and Vijay 1994). Antimicrobial activity of sweet basil has been found at least against such organisms as *Lactobacillus acidophilus*, *Saccharomyces cerevisiae*, *Mycoderma* sp., *Aspergillus niger*, *Bacillus cereus* (Meena and Vijay 1994), *Staphylococcus aureus*, *E. coli*, *Candida albicans*, *Corynebacterium* sp. (Morris *et al.*, 1979) and even against *Penicillium italicum* (Rewa and Pandey 1977), which is a common mold in citrus fruits. Eugenol is reported to have an inhibitory effect on *Bacillus subtilis*, *Salmonella enteritidis*, *Staphylococcus aureus*, *Pseudomonas aeruginosa*, *Proteus vulgaris* and *E. coli* (Katayama and Nagai 1960).

IS USE OF BASIL MICROBIOLOGICALLY SAFE?

When herb plants are used as such, dried or ground, there is a danger that they are contaminated by microbes and microbial toxins. Hartgen and Kahlau (1985) investigated 719 samples of 25 commonly used spices collected from the retail trade. Clostridial contamination was detected in 20%; mainly in leaf herbs like basil and coriander. Altogether 108 strains of clostridia were isolated, of which 38 were strains of *Cl. perfringens*, one of the common causes for food poisoning. Aflatoxins have been found in basil. Of a total of 72 samples of traditional herbal drugs from retail stores in Indonesia, however, only two samples of basil were found to contain aflatoxin (Tanaka *et al.*, 1988).

DRYING HERBS

In connection with the Finnish study of herbs (Mäkinen *et al.*, 1986) Malmsten *et al.* (1991) tested the microbiological quality of sweet basil (Table 6.1). Aerobic sporeformers were present from 200 to 3.5 × 10^3/g dried basil. *Bacillus cereus* was present in all samples, counts ranging from 50 to 2.2 × 10^3/g dried basil. These levels are not high enough to cause food poisoning. Anaerobic sporeformers were randomly distributed, and counts were very low; from 5 to 30 organisms/g dried herb. After one year storage, the APC (aerobic plate count) of the air-dried basil averaged ten times higher than that of the freeze-dried basil. It was concluded that more microbes were destroyed by freeze-drying than by air-drying. Coliforms and fecal streptococci were found in both freeze-dried and air-dried samples, but only sporadically and at very low counts. The method of drying had no significant effect on the content of fecal streptococci. Molds and yeasts were found in almost all samples, and they were approx. ten times higher in air-dried than in freeze-dried samples. Survival of molds and yeasts depended on the presence of oxygen in packages (Table 6.1). Presence of the aerobic sporeformers did not depend on the types of packages. Storing temperature affected the survival of microbes. The counts, especially those of yeasts and molds, were higher from the samples stored at 23°C than at 35°C. The higher temperature (35°C) was evidently unfavorable for coliforms and for fecal streptococci as well, and *E. coli* was not detected in the samples stored at 35°C. Consequently the changes in sensory quality of the samples stored at 35°C were due to the

Table 6.1 Microbial counts of freeze-dried and air-dried basil after storage at 23°C (Results of counts on five replicates) (Malmsten *et al.*, 1991)

		Storage Time **1 Year**			
Process		*Freeze-dried*		*Air-dried*	
	Glass Jar	*Vacuum*	*Nitrogen*	*Glass Jar*	*All Samples*
Microbial Group		*Geometric Mean*			*Range*
APC	5 × 10^4	8 × 10^4	9 × 10^4	7 × 10^5	2.2 × 10^4 – 5.9 × 10^6
Coliforms	< 3	6	< 3	17	< 3 – 1.1 × 10^4
E. coli	< 3	< 3	< 3	< 3	< 3 – 9
Fecal streptococci	< 10	< 10	< 10	12	< 10 – 440
Molds and yeasts	1 × 10^3	3 × 10^3	4 × 10^3	2 × 10^4	200 – 8.7 × 10^4
Aerob. sporeform.	1 × 10^3	2 × 10^3	1 × 10^3	2 × 10^3	650 – 3.5 × 10^3
Bacillus cereus	920	1 × 10^3	660	750	50 – 2.2 × 10^3
Clostridia	< 10	< 10	< 10	< 10	< 10 – 30
		Storage Time **2 Year**			
APC	3 × 10^4	4 × 10^5	3 × 10^5	4 × 10^4	3.2 × 10^4 – 3.0 × 10^6
Coliforms	< 3	< 3	< 3	13	< 3 – 240
E. coli	< 3	< 3	< 3	< 3	< 3
Fecal streptococci	< 10	< 10	< 10	12	< 10 – 120
Molds and yeasts	490	1 × 10^4	6 × 10^3	2 × 10^3	60 – 2.5 × 10^4
Aerob. sporeform.	850	490	730	760	200 – 1.7 × 10^3
Bacillus cereus	310	450	510	970	100 – 1.6 × 10^3
Clostridia	< 10	< 10	< 10	< 10	< 10

chemical changes and not due to microbiological deterioration (Pääkkönen *et al.*, 1990). The essential factor for determining the microbiological quality of dried basil was the quality of the fresh material. Nevertheless, the drying method, type of packaging, and storage conditions also have an effect on the microbiological quality of the herb (Malmsten *et al.*, 1991).

HERBS CAN BE PRESERVED BY IRRADIATION

Spices and herbs have been treated with ethylene dioxide in order to reduce their germ and mold content, but the method is now forbidden in most countries. Therefore ionization methods have been developed to reduce and to prevent microbial contamination. Many consumers, however, are very much against irradiation of food, and if done, it has to be mentioned in food labelling. Food chemists are trying to find a simple method to detect the irradiation of herbs and spices. Thermoluminescence analysis has been found to be 99% successful in detecting irradiated herbs (Schreiber *et al.*, 1995). Even one year after irradiation, basil showed even greater luminescence intensity (Heide and Boegl 1987). However, according to Heide and Boegl (1985), exposure of the herb to moisture subsequent to irradiation makes the ionization treatment undetectable.

DRYING CONDITIONS DETERMINE THE SENSORY QUALITY OF DRIED PRODUCT

To enhance preservation, herb plants are subjected to various treatments, including hot air drying, microwave drying, infrared drying, and freeze-drying. In hot air drying, temperature, air velocity and humidity are specific to each plant. Undesirable effects may occur, *e.g.* loss of aromatic substances and change in color. According to Baritaux *et al.* (1991) the best candidate for the enzymatic browning reaction in basil is probably rosmarinic acid, the most abundant phenol in fresh basil leaves. One of the causes for the sensitivity of basil to enzymatic browning is also the weak acidity of basil sap, its pH being only 5.6.

Drying of basil reduces the growth of microorganisms and prevents some biochemical reactions. Drying, however, gives rise to a number of negative physical and chemical modifications of the quality of basil due to changes in appearance, taste and smell. The rate of drying is controlled mainly by the temperature. Rocha *et al.* (1992) have studied drying of mint and basil. Blanching strongly reduces the drying time. If basil is not blanched, the changes in the color due to the degradation of chlorophyll a and b and of carotenoid pigments as well as enzymatic and nonenzymatic browning are better prevented when the drying is done at low temperature (45–50°C). A short steam blanching inactivated the enzymatic browning and inhibited the degradation of the carotenoid and the chlorophyll pigments as well. When basil is blanched for 15 sec before drying, the drying rate is ten times quicker than without blanching treatment. When basil was dried at 50°C (Rocha *et al.*, 1993), the drying behavior differed from that at 35, 40, and 45°C, probably due to the melting of the fine wax layer on the leaves, thus allowing a faster water migration.

Baritaux *et al.* (1992) have studied drying and storage of basil (*Ocimum basilicum* L.). In this study, fresh basil with a water content of 86.5% w/w was dried at 45°C for 12 hours until the final water content was 10% w/w; thereafter it was stored in aluminum polyethylene bags at 4°C for 3, 6 and 7 months. The samples were submitted to steam distillation, and the components of the oil were analyzed and compared with those of fresh basil oil samples. The losses of total essential oil after drying were 19%, 62% and 66% after storage of 3, 6, and 7 months, respectively. A significant decrease in the content of methylchavicol and eugenol was mainly responsible for the overall loss of essential oil. In order to prevent the loss of volatile compounds in basil oil, drying under 50°C, preferably not above 40°C, is recommended (Deans *et al.*, 1991).

BASIL OIL CAN BE STORED IN MICROCAPSULES

When basil oil is stored in microcapsules, its storage life is increased, its use in food industry is safer, and no hygienic problems result due to microbial contamination. According to Sheen *et al.* (1992) basil microcapsules were more stable than those of garlic or ginger with regards to oil retention. Most essential oil was retained, when the microcapsule composed of soy meal and maltodextrin, 5% w/w and 50% w/w respectively. Double-walled microcapsules provided the slowest dissolution rate and the greatest oil stability (Sheen and Tsai 1991). Marion *et al.* (1994) have reported a method to encapsulate herbs into an oleoresin complex of gelatine and natural gums without emulsifiers. The water solubility in this capsule was perfect, sensory properties excellent and the product was microbiologically safe.

A basil flavoring has been developed that can be directly sprayed to foods. The extract of basil is centrifuged from ground material and stored in bottles under pressure. The method is patented in Italy (Antonucci-Tarolla 1988). Herb concentrates add the authentic flavor of fresh herbs to foods, and in this way product developers can get a "fresh taste" with herb concentrates. They are particularly suited to microwave foods, where instant flavor formation is necessary due to the very short cooking time (Anon. 1992).

Extracts, flavorings, oils, spice and oleoresin of sweet basil, *Ocimum basilicum*, are on the American Food and Drug Administration's GRAS list, which means that they are generally recognized as safe. Spice and flavorings of *Ocimum minimum* are on the list as well (McCaleb 1994).

BASIL AND BASIL OIL ARE USED IN MANY FOODSTUFFS

Meat and Fish

Due to its antimicrobial properties and antioxidative effects, basil oil is used to increase the shelf life of many foodstuffs. In Russia there are patents pertaining to the addition of comminuted basil or basil oil in the manufacture of sausages and other meat products (Guseinov *et al.*, 1992; Madaliev *et al.*, 1990, Mamedov *et al.*, 1983). The addition of basil oil results in better flavor and color intensity and in less microbial contamination (Mamedov *et al.*, 1984, Dinarieva *et al.*, 1984). An amount of only 5–10% w/w of basil in a mixture of other spices in sausage products improved the organoleptic characteristics of the products (Dinarieva *et al.*, 1982). Germans are also adding herbs in sausages (Anon 1988). The flavor of sausages was retained better, when basil was added in the form of micro-encapsulated oleoresin, which penetrated even the intramuscular fats in the sausages (Flint and Seal 1985).

According to Polic and Nedeljkovic (1978) basil causes off-taste in poultry meat. However, a Lebanese traditional dish, stuffed turkey, contains basil as an ingredient (Salah 1977), and there are recipes for poultry dishes with basil in many cookbooks. Reuter (1976) has written several review articles on the origin, properties and applications of the most important spices used for meat products, including basil. Basil is recommended to flavor beef stews and bouillons, pork, meat loaf, meat balls and shish kebabs, lamb dishes and inner organ dishes, as well as chicken and turkey. Canned beef stew tastes homemade when basil is added. Basil suits all fish dishes, including stuffings, and it may also be used in other seafood dishes.

Butter, Cheese, Sauces and Dressings

The Italian firm Yomo has developed a new "fresh table dairy speciality", which it is marketing under the name "**Belgioioso**", a readily spreadable creamy product which contains 18.5 g fat, 9 g protein, 3 g carbohydrate (905 kJ; 216 kcal/100 g) with typical variations of Mediterranean flavors, including basil (Maiocchi 1995). Spiced butter gets also a pleasant flavor from extract of basil, and it is popular in France, where it is patented (Laiteries *et al.*, 1971). Basil or basil oil has been added to cheeses, for example "**Brodinskii**", a processed cheese made in Russia (Samodurov *et al.*, 1991), and a low-calorie, low-sodium, low-cholesterol cheese named "**Pre Monde**", a cheddar-type cheese in the USA (Anon 1984b), which is a condiment consisting of dried sweet basil and cheese (Biller and Kellerman 1980).

Many sauces used to enhance entrees and side-dish food products contain basil; **Pesto** is one of the most popular, and its basic ingredients are garlic, basil, and olive oil. It is used with pasta dishes, soups, dips etc. Pesto originates from Genoa, Italy. It can be prepared in home cooking, but it is also commercially available in different varieties, and it is especially popular in Mediterranean countries (Dziezak 1991, Anon 1984a, Anon 1981). Due to its herb ingredients pesto is shelf stable and is thus said to be unique (Ranch 1990). Cookbooks contain many recipes for basil-containing dressings and sauces, including basil honey, basil yogurt, basil tomato, basil vinaigrette, basil mustard, basil anchovies, and basil Parmesan sauce.

Basil in Beverages

Many alcoholic beverages, for example bitters, liquors and spirits, contain basil. **Chartreuse** liquor owes its fine fragrance partly to basil. Russians have patented a method to improve the storage stability and organoleptic properties of a carbonated fermented milk beverage by adding a mixture of essential oils of coriander, basil and fennel to a salt solution of whey (Askerova *et al.*, 1993). A non-alcoholic beverage called **Reikhan** is also patented in Russia. It is made by an infusion of basil leaves and stems in water at 95–100°C. It is filtered, mixed with citric acid and sugar, and then cooled (Kerimov 1993). Russians have patented a method to prepare a non-alcoholic beverage concentrate even from the residues of grapes and high-eugenol basil (Bagaturiya *et al.*, 1991). Based on a German patent, basil (1–40 g/l), fresh, dried or frozen, is also used in sweet or dry alcoholic beverages based on spirits, garlic or lemon (Meier 1990). Instead of basil leaves, basil seeds are used as an ingredient of alcoholic liquors like **Dushanbinskaya** bitters (Bagdasarov *et al.*, 1978). In Ukraine, screening of 20 wild-growing aromatic plants led to development of four new soft drinks based on wormwood, mint, fennel, borage and basil. These beverages have excellent sensory properties, contain significant amounts of vitamin C, thiamin, and riboflavin. Due to their antimicrobial substances derived from herbs, the beverages have a shelf life of 10–12 days (Kolesnikova *et al.*, 1976).

Other Uses of Basil in Foodstuffs and in Foods

Sweet basil or its extracts and oils are widely used in combination with other spices and herbs in confectionary products, sweets, bakery products, puddings, condiments, vinegars, ice creams, mustard and in pickled vegetables. Basil is a common ingredient in spice mixtures. Russians have patented a basil-containing salt substitute without sodium salts for dietetic use (Rakauskas and Sudzhene 1979). Basil leaf flavoring is patented in Great Britain (Ricci 1974) and in Germany (Huth and Werner 1973). In 1974 Norda Inc. in New York introduced a product called **Spice Grains** in which a herb oil extracted from natural herbs like basil was incorporated into a carrier of cereal solids with mono- and diglyceride antioxidants and food coloring. This product is used in candy and in the snack industry (Anon 1974).

Basil in Home Cooking

Herbs are used mostly as fresh in those countries which grow them traditionally; their aroma and flavor are then definitely at their best. A fresh herb has a smoother flavor than a dried one. Herbs suited well to be used with basil are sage, garlic, common balm, oregano, parsley, thyme, rosemary and summer savory.

Rice noodles and vegetables get a fine aroma when cooked in herb liquids. The liquid is prepared by simmering herbs in water. If you prepare a herb drink, for example basil tea, brewing time is at least half an hour. Two and a half centiliters of boiling water is poured on two tablespoons of fresh basil or on one tablespoon of dried basil. You can also season noodles and rice, omelets and soups with sweet basil. Turtle soup, oxtail soup, and the French soup **paysanne** contain basil as an essential ingredient. Basil goes well also with cheese and cheese dishes.

Potatoes, tomatoes, spinach, cabbage, beans, zucchini, squash, aubergine, mushrooms, lettuce, artichokes, broccoli, brussel sprouts and carrots are vegetables often flavored with basil. Serving basil in pea and bean dishes reduces the flatulence caused by the indigestible sugars in these legumes.

Desserts and home-baked products can also be flavored with basil, which goes well with apples, apple pie, apple sauces and apple jellies, for example. Cookbooks give many recipes for basil biscuits, basil potato scones, bread with basil, basil souffles, etc.

Sheen and Tsai (1991) have analyzed flavor characteristics of sweet basil by sensory evaluation. Oils obtained from the various parts of the basil plant were categorized into five groups by preference ranking test for their aromas. Among them the aroma of the fresh leaf-flower oil was the most preferable, while that of the stem was the least one. According to odor assessment, methyl-chavicol had the most preferable odor, and 1-octen-3-ol the least favorable one. Consequently the European, the Reunion "exotic" and the Bulgarian basil have the most popular odors, *e.g.* they are rich in methyl-chavicol.

Basil, like other herbs, can be used in preparing a variety of foods. Law (1975) has written a book on the use of herbs in cooking with recipes for sandwiches, omelets, soups, biscuits, salads and how to use herbs in garnishing meat, fish and poultry. Basil is one of the more than 40 herbs listed.

Principles of Seasoning with Herbs

Although herbs and spices are not the main ingredients of our food, they are important in flavoring and enhancing the flavor and smell of many dishes. "Add spice/herb according to taste" is often written in cookbooks, and indeed, it is nearly impossible to tell the exact amount of each spice or herb for a certain dish. However, there are some general rules to be followed in order to receive the best result with herb flavors.

Dressings and stuffings, which are used to add to the flavor of the basic foodstuff, need more herbs and/or spices than soups, which are eaten as such. A mild-tasting steamed fish is not flavored as strongly as a strong meat stew, and a cold dish always needs more spices or herbs than a warm dish, and a frozen one the most. A pinch of salt or sugar may enhance the flavor of the herb.

When cooking with fresh herbs, add them just before serving. Their aromatic substances do not stand high temperatures; they are easily evaporated, destroyed or changed in flavor. If you have to use dried herb instead of fresh one, two teaspoons of a fresh herb correspond 1/4 teaspoon of a dried one.

Preserving Basil at Home

Basil can be processed and stored at home by drying or freezing or storing in oil, vinegar, butter, mayonnaise or in sauces, for example in homemade pesto. Basil bundles are tied loosely and hung upside down in a cool, dark, well ventilated place for several weeks. When the leaves are completely dry, they should be stripped from the stem and stored in a dark-walled, air-tight container in a cool place.

It is possible to preserve basil also by drying in a microwave oven. Sprigs of basil are placed between paper towels, put in a microwave oven and heated with full power until the leaves are completely dry, e.g. brittle. Then they are stored as described above.

When freezing basil, the sprigs should be rinsed and patted dry. The leaves are stripped from the stems and put into a plastic bag, which is gently flattened to remove the air and sealed tightly. It is more convenient to grate the herb when frozen rather than melt it. Basil can also be "preserved" in oil, butter, mayonnaise, salad dressing, vinegar and wine. It should be shredded rather than chopped, and then added to the foodstuff at least one hour, preferably one day before serving. Vinegar with herbs must be made from wine. Whole sprigs can also be added to oil or vinegar. Many cookbooks say that oil or vinegar does not get its full flavor from herbs until after several weeks.

Do not Store Fresh Basil in Refrigerator

Lange and Cameron (1994) have shown that the best temperature for freshly harvested greenhouse sweet basil is 15°C. When packed in low-density polyethylene film covered with black plastic material, the shelf life of basil was 12 days. Chilling injury symptoms, e.g. darkening of the leaves, were severe already at 5°C, and the shelf life was only 3 days. At 0°C basil could be stored only for one day. Consequently, a refrigerator is not a good place to store fresh basil.

THE FINNISH HERB STUDY

Sweet basil was one of the herbal plants included in a large Finnish herb study in 1983–85 (Mäkinen *et al.*, 1986). The research institutes participating in the study were: Department of Horticulture, Department of Food Chemistry and Technology, Department of Agricultural Economics and Department of Nutrition of Helsinki University; Department of Chemistry and Biochemistry of Turku University, Technical Research Centre of Finland and Research Laboratories of the State Alcohol Company.

The basil grown in southern Finland belongs to the linalool-estragol chemotype, and the estragol content of its oil is clearly higher than the linalool content. A total of 54 different compounds were identified from the essential oil of sweet basil cultivated in Finland; seven of the compounds being not reported earlier (Nykänen 1986). According to Hälvä (1987) the volatile oil of a dried sample of basil harvested at time of bud formation changed from 0.46 to 0.93% w/w when analyzed.

A fresh herb has a smoother flavor than a dried one. The water content of fresh basil is 85% w/w compared to 8% w/w of air dried basil and 3% w/w of freeze-dried basil (Pääkkönen 1986). Pääkkönen *et al.* (1990) studied the effects of drying and packaging on the quality of sweet basil. The intensity scoring of odor and taste (Figure 6.1) was higher for dried than for the fresh basil, although the odor of the fresh basil changed drastically due to drying. The taste of air-dried basil did not differ significantly from that of the frozen or freeze-dried basil. Przezdziecka and Baldwin (1971) have made sensory evaluation experiments with herbs and spices and found out that the threshold intensity of taste and aroma of basil is "intermediate", when compared with that of garlic ("high") and marjoram ("low").

In the Finnish herb study Pääkkönen's *et al.* (1990) sorption isotherms showed that herbs dried with different methods exhibited different sorption capacities and hysteresis. Above 0.5 a_w the freeze-dried sample showed a negative hysteresis effect (Figure 6.2a,b) similar to the sorption isotherm of freeze-dried dill (Pääkkönen *et al.*, 1989). The sorption hysteresis may be due to the changes occurring in the physical structure of the herb plant during the drying process. The quality of basil products

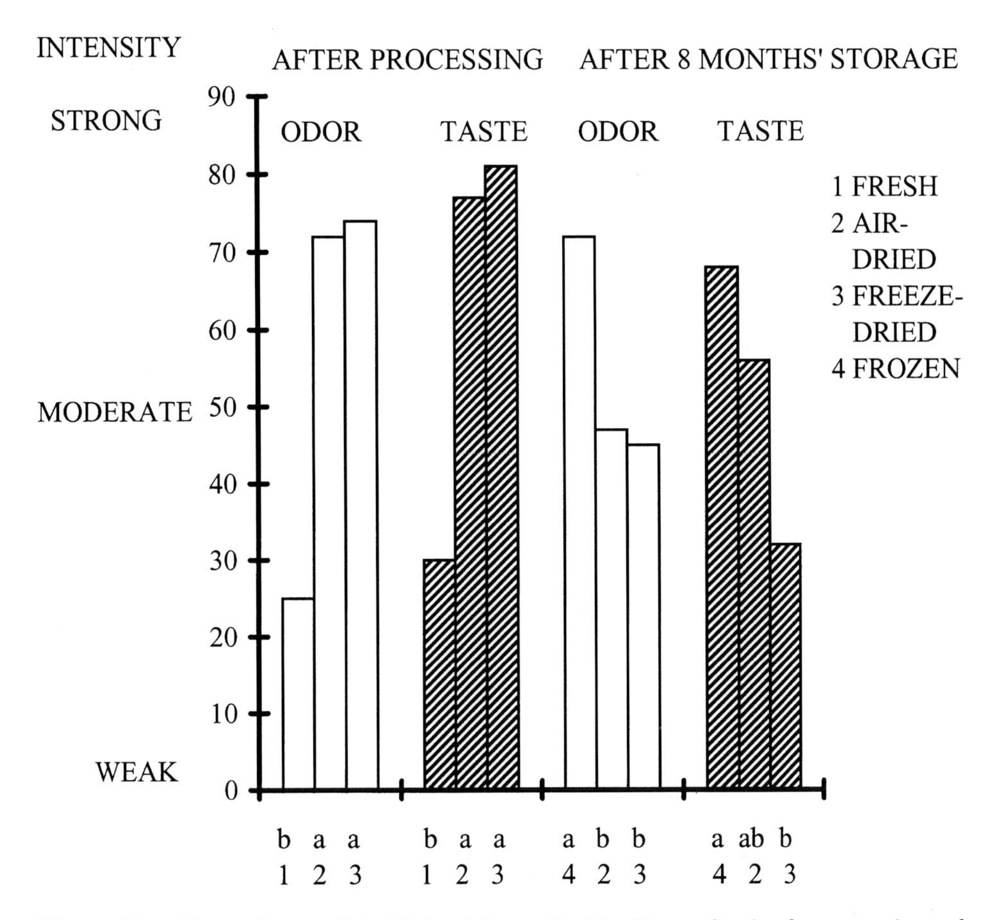

Figure 6.1 Odor and taste of air-dried and freeze-dried basil immediately after processing and after 8 months in glass jars at room temperature compared with fresh basil and frozen basil stored in a freezer for 8 months. Means with unlike superscripts are significantly different at the 99% level using Tukey's test.
0 = weak; 100 = strong intensity (Pääkkönen *et al.*, 1990).

was influenced by the methods of drying and packaging. The color of basil was affected more by the drying process than by the storage conditions. Freeze-dried basil exhibited an intensive green color compared to the brown color of air-dried basil; nine months' storage in light at room temperature or in the dark at 35°C altered the color of the freeze-dried basil only slightly. The odor and taste of freeze-dried basil were very well preserved when stored at room temperature either in vacuum package or packed in nitrogen atmosphere or even in dark glass jars. At room temperature two years' storage in dark glass jars did not cause any significant weakening of the odor or taste of either air-dried or freeze-dried basil. However, it is better to pack freeze-dried basil in vacuum or under nitrogen than in paper bags, but packaging is of no importance for air-dried basil. At any rate, elevated storage temperature (35°C) affects the sensory quality of dried basil more than the method of

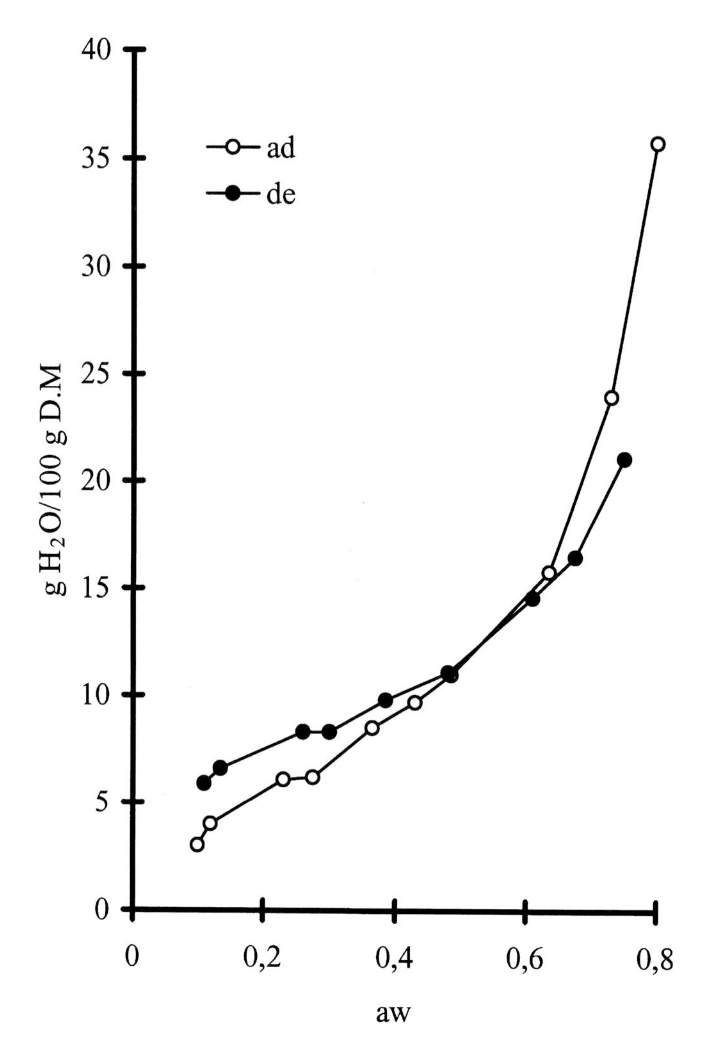

Figure 6.2a Sorption isotherms of freeze-dried basil at 23°C. ad = adsorption; de = desorption (Pääkkönen *et al.*, 1990).

packaging. At room temperature the intensity and taste of dried basil in an air-tight package was maintained even for two years.

Cultivation, chemical composition, processing, packaging and marketing of herbal plants in addition to basil such as coriander, cumin and mustard (seed herbs); dill, sweet and wild marjoram (leaf herbs) and garden angelica (root herb) were investigated in this large Finnish herb study in 1983–85 (Mäkinen *et al.*, 1986). Four Ph.D dissertations, several master theses in addition to many scientific publications have been published within the framework of this study. The amounts and the quality of the crops, processing *e.g.* drying processes after harvesting, packaging and storage conditions and the possibilities to use the herbs cultivated in Finland in the food industry and in catering were investigated and evaluated. It is worthwhile pointing

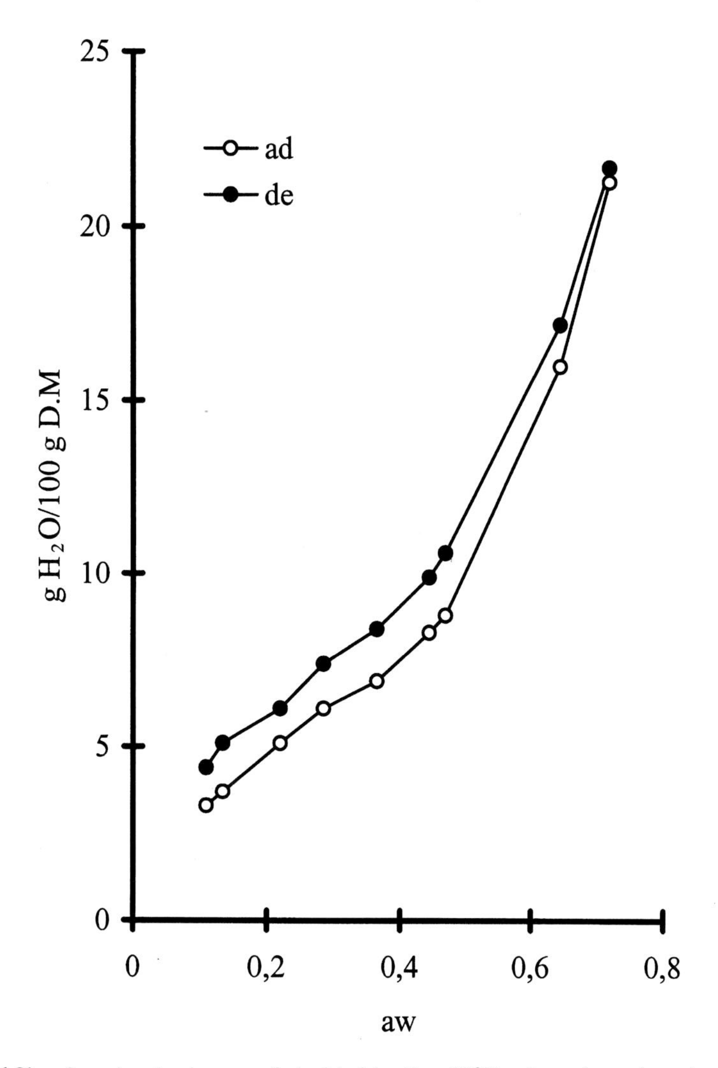

Figure 6.2b Sorption isotherms of air-dried basil at 23°C. ad = adsorption; de = desorption (Pääkkönen *et al.*, 1990).

out that all the determinations were made, though in different laboratories, from the same crops. Consequently this research has established how to grow, fertilize, harvest, dry and pack these herbs in Finland.

REFERENCES

Anon (1974) Oils of spices used to create new flavour material. *Candy Ind.*, **139**, 45.

Anon (1981) Herb sauce offers new flavour for pasta. *Food Development*, **15**, 26–29.

Anon a (1984) All it takes is water and oil to make Instant Pesto. *Food Eng.*, **56**, 56.

Anon b (1984) Cheese maker supports users to keep name visible. *Quick frozen foods*, **46**, 36–37.
Anon (1988) Herbs for sausage manufacture. *Fleischerei*, **39**, 808–810.
Anon (1992) Product developers get 'fresh' with herb concentrates. *Prepared Foods*, **161**, 129.
Antonucci-Tarolla, M. (1988) System for aromatizing foods. European Patent Application, EP0259274A1.
Aruna, K. and Sivaramakrishnan, V. (1992) Anticarcinogenic effects of some Indian plant products. *Food Chem. Toxicol*, **30**, 953–956.
Askerova, A., Guseinov, I., Azimov, A., Dmitrieva, N., Shamsizade, R. (1993) Manufacture of the carbonated fermented milk beverage, Airan. USSR Patent, SU1796122.
Bagaturiya, N., Mekhuzla, N., Tabatadze, T. and Kondzhariya, L. (1991) Manufacture of a concentrate for non-alcoholic beverage. USSR Patent, SU1671239A1.
Bagdasarov, S., Nazyrova, R., Airopet'yants, B. and Voitenko, T. (1978) Composition of ingredients for "Dushanbinskaya" bitter lemon. USSR Patent, 587151.
Baritaux, O., Amiot, M., Richard, H. and Nicolas, J. (1991) Enzymatic browning of basil (*Ocimum basilicum* L.). Studies on phenolic compounds and polyphenol oxidase. *Sciences des Aliments*, **11**, 49–62.
Baritaux, O., Richard, H., Touche, J. and Derbesy, M. (1992) Effects of drying and storage of herbs and spices on the essential oil. Part I. Basil (*Ocimum basilicum* L.). *Flavour and Fragrance Journal*, **7**, 267–271.
Bassiouny, S., Hassanien, F., Abd El Razik, A. and El Kayati, M. (1990) Efficiency of antioxidants from natural sources in bakery products. *Food Chem.*, **37**, 297–305.
Beckstrom-Sternberg, S. and Duke, J. (1994) Potential for synergistic action of phytochemical in spices. *J. Food Sci. Technol. Ind.*, **31**, 68–70.
Bersani, C., Soncini, G. and Cantoni, C. (1981) Evaluation of mutagenic activity in essences and spices by the Ames test. *Arch. Vet. Ital.*, **32**, 10–11.
Biller, F. and Kellerman, R. (1980) German Federal Republic Patent Application, 2924358.
Charalambous, G. (ed.) (1994) *Spices, Herbs and Edible Fungi*, Elsevier Science B.V., The Netherlands.
Chen, B., Chuang, J., Lin, J. and Chiu, C. (1993) Quantification of provitamin A compounds in Chinese vegetables by high-performance liquid chromatography. *J. Food Prot.*, **56**, 51–54.
Deans, S., Svoboda, K. and Bartlett, M. (1991) Effect of microwave oven and warm-air drying on the microflora and volatile oil profile of culinary herbs. *Journal of Essential Oil Research*, **3**, 341–347.
Dinarieva, G., Solntseva, G., Belusova, E., Bushkova, L., Klenina, Z., Pruidze, V. and Tukvadze, S. (1982) Composition of spices for sausages. USSR Patent, 921491.
Dinarieva, G., Solntseva, G., Belusova, E., Kandilov, N., Shukyurov, N., Kocharin, A. and Guseinov, V. (1984) Spice essential oil composition for use in sausage production. USSR Patent, SU1069755A.
Dziezak, J. (1991) Getting savvy on sauces. *Food Technol.*, **45**, 84, 87.
Felice, M., Leonardis, T. and Comes, S. (1993) Effects on autoxidation. *Industrie Aliment*, **32**, 249–253.
Flint, F. and Seal, R. (1985) The sausage seasoning scene. *Food Manuf.*, **60**, 43, 45.
Fomicheva, L., Keller, E., Gulyaev, V., Roenko, T. and Koptyaeva, I. (1982) Activity of vegetable additives used for food concentrates. *Konservnaya i Ovoshchesushil' naya promyshlennost'*, **1**, 39–41.
Friedman, M. (1993) New blends/ethnic varietes are the life of spice. *Prepared Foods*, **162**, 59.
Guseinov, V., Shukyurov, N., Movsum-zade, A., Ibragimov, F., Sadykhova, R., Askerova, K. and Alieva, D. (1992) Manufacture of sausages. USSR Patent, SU1722371.
Hälvä, S. (1987) Studies on fertilization of dill (*Anethum graveolens* L.) and basil (*Ocimum basilicum* L.). III Oil yield of basil affected by fertilization. *J. Agric. Sci. Finland*, **59**, 25–29.
Hartgen, H. and Kahlau, D. (1985) Colony counts in spices used in the home. *Fleischwirtschaft*, **65**, 99–102.
Heide, L. and Boegl, W. (1985) Chemiluminescence measurements on 20 types of herbs and spices. Method for detection of treatment with ionizing radiation. *Z Lebens-Unters Forsch*, **181**, 283, 288.

Heide, L. and Boegl, W. (1987) Identification of irradiated spices with thermo- and chemiluminescence measurements. *Int. J. Food Sci. Technol.*, **22**, 93–103.

Holland, B., Unwin, I. and Buss, D. (eds.) (1991) in McCance & Widdowson's *The Composition of Foods. Vegetables, Herbs and Spices*, Bath, U.K.

Huth, H. and Werner, K. (1973) Method of obtaining water-soluble, practically germfree flavourings from fresh herbs. German Federal Republic Patent Application, 2208815.

Katayama, T. and Nagai, I. (1960) Bull Jpn Soc Sci Fisheries. Ref. Nakatani, N. Antioxidative and Antimicrobial Constituents of Herbs and Spices. In: G. Charalambous (ed.) 1994, *Spices, Herbs and Edible Fungi*. Elsevier Science B.V., The Netherlands, pp. 251–271

Kerimov, T. (1993) Manufacture of the non-alcoholic beverage Reikhan. USSR Patent, SU1796122.

Klaui, H. (1973) Natural occurring antioxidants. Proceedings of the Institute of Food Science and Technology of the U.K., **64**, 195–209. Ref. Tsimidou, M., Boskou, D. Antioxidant Activity of Essential Oils from the Plants of the *Lamiacae* Family. In: G. Charalambous (ed.) 1994, *Spices, Herbs and Edible Fungi.* Elsevier Science B.V., The Netherlands, pp. 273–284.

Kolesnikova, I., Nilov, G., Baranova, S., Ermakov, A. and Pirog, N. (1976) New plant raw materials for production of soft drinks. *Kharchova Promislovist*, **6**, 43–44.

Laiteries, E. and Bridel, S. (1971) Spiced butter. French Patent Application, 2060213.

Lange, D. and Cameron, A. (1994) Postharvest shelf life of sweet basil (*Ocimum basilicum*). *Hortic. Sci.*, **29**, 102–103.

Law. (1975) *Herbs for Health and Flavour*. John Bartholomew & Son Ltd. Edinburgh. U.K. 62 pp.

Madaliev, I., Zharinov, A., Fazylov, A. and Safutdinova, D. (1990) Production of semi-smoked sausages. USSR Patent.

Maftei, M. (1992) Prospects in European market for culinary herbs. *Int. Trade Forum.*, **1**, 4–9, 34.

Maiocchi, G. (1995) Ultraspreadable. *Latte*, **20**, 490–493.

Mäkinen, S., Hälvä, S., Pääkkönen, K., Huopalahti, R., Hirvi, T., Ollila, P., Nykänen, I. and Nykänen, L. (1986) *Maustekasvitutkimus*. The Academy of Finland, Report SA 01/813, Helsinki, Finland.

Malmsten, T., Pääkkönen, K. and Hyvönen, L. (1991) Packaging and storage effects on microbiological quality of dried herbs. *J. Food Sci.*, **56**, 873–875.

Mamedov, A., Guseinov, V. and Shukyurov, N. (1983) Use of domestic Caucasian spices to flavour meat products. *Proc. Eur. Meet Meat Res. Workers*, **29**, 525–529.

Mamedov, A., Guseinov, V., Aliev, S. and Shukyurov, N. (1984) Effect of essential oils on the quality of semi-smoked sausages during storage. *Proc. Eur. Meet Meat Res. Workers*, **30**, 350–352.

Marion, J., Audrin, A., Maignial, L. and Brevard, H. (1994) Spices and Their Extracts: Utilization, Selection, Quality Control and New Developments. In G. Charalambous (ed.) 1994, *Spices, Herbs and Edible Fungi*, Elsevier Science, B.V., The Netherlands, pp. 71–95.

Mathews, S., Singhal, R. and Kulkarni, P. (1993) *Ocimum basilicum*: a new non-conventional source of fibre. *Food Chem.*, **47**, 399–401.

McCaleb. (1994) Food Ingredient Safety Evaluation. In. G. Charalambous (ed.) 1994, *Spices, Herbs and Edible Fungi*, Elsevier Science, B.V., The Netherlands, pp. 131–143.

Meena, M. and Vijay, S. (1994) Antimicrobial activity of essential oils from spices. *J. Food Sci. Technol. (Calcutta)*, **31**, 68–70.

Meier, G. (1990) Alcoholic beverage. German Federal Republic Patent Application, DE3833937A1.

Morris, J., Khettry, A. and Seitz, E. (1979) *J. Am. Oil Chem. Soc.*, **56**, 595–603, Ref. Nakatani, N. Antioxidative and Antimicrobial Constituents of Herbs and Spices. In G. Charalambous (ed.) 1994, *Spices, Herbs and Edible Fungi*, Elsevier Science B.V., The Netherlands, p. 265.

Nykänen, I. (1986) High resolution gas chromatographic-mass-spectrometric determination of the flavour composition of basil (*Ocimum basilicum* L.) cultivated in Finland. *Z Lebens-Unters Forsch*, **182**, 205–211.

Pääkkönen, K. (1986) Sadon käsittely ja aistinvarainen laatu. In S. Mäkinen, S. Hälvä, K. Pääkkönen, R. Huopalahti, T. Hirvi, P. Ollila, I. Nykänen, L. Nykänen. *Maustekasvitutkimus*. The Academy of Finland, Report SA 01/813, Helsinki, Finland, pp. 40–65.

Pääkkönen, K., Malmsten, T. and Hyvönen, L. (1989) Effects of drying method, packaging and storage temperature and time on the quality of dill (*Anethum graveolens* L.) *J. Food Sci.*, **54**, 1485–1487, 1495.

Pääkkönen, K., Malmsten, T. and Hyvönen, L. (1990) Drying, packaging and storage effects on quality of basil, marjoram and wild marjoram. *J. Food. Sci.*, **55**, 1373–1377, 1382.

Polic, M. and Nedeljkovic, L. (1978). Effect of addition of spices on the flavour of chicken. *Tehnoligija Mesa*, **19**, 359–362.

Przezdziecka, T. and Baldwin, Z. (1971) Sensory characterization of spices. *Roczniki Instytutu Przemyslu Miesnego*, **8**, 45–46.

Rakauskas, A. and Sudzhene, S. (1979) Salt substitute for dietetic use. USSR Patent, 648196.

Ranch, C. (1990) Particle size key to new Italian pesto product. *Food Rev.*, **17**, 43, 46.

Reuter, H. (1976) Die wichtigsten Gewurze fur Fleischereien und Gaststätten. *Fleischwirtschaft*, **56**, 188, 191–192, 296, 489–490, 639–641, 812–813.

Rewa, A. and Pandey, G. (1977) The application of essential oils and their isolates for blue mould decay control in Citrus reticulata Blanco. *J. Food Sci. Technol. (Calcutta)*, **14**, 14–16.

Ricci, A. (1974) Basil leaf flavouring. British Patent, 1348266.

Rocha, T., Lebert, A. and Marty-Audouin, C. (1992) Effect of Drying Conditions and of Blanching on Drying Kinetics and Color on Mint (*Mentha spicata Huds.*) and Basil (*Ocimum basilicum*). In A. Mujumdar (ed.) **Drying**. Elsevier Science Publishers B.V. The Netherlands, pp. 1360–1369.

Rocha, T., Lebert, A. and Marty-Audouin, C. (1993) Effect of pretreatments and drying conditions on drying rate and colour retention of basil (*Ocimum basilicum*). *Lebensm-Wiss Technol.*, **26**, 456–463.

Saito, Y. and Asari, T. (1976) Studies on the antioxidant properties of spices. Total tocopherol content in spices. *J. Jpn. Soc. Food Nutr.*, **29**, 289–292.

Saito, Y., Kimura, Y. and Sakamoto, T. (1976) Studies on the antioxidant properties of spices. III. The antioxidant effects of petroleum ether soluble and insoluble fractions from spices. *J. Jpn. Soc. Food Nutr.*, **29**, 505–510.

Salah, N. (1977) *Arab World Cook Book*. Tien Wah Press (Pte) Limited, Singapore.

Samodurov, V., Dolgoshchinova, V., Tarasyuk, V., Pruidze, G., Kraevaya, N. and Yakhnev, N. (1991) Composition for manufacture of 'Brodinskii' processed cheese. USSR Patent, SU1690657.

Schreiber, G., Helle, N. and Boegl, K. (1995) An interlaboratory trial on the identification of irradiated spices, herbs and spice-herb mixtures thermoluminescence analysis, *JAOAC Int*, **78**, 88–93.

Sheen, L., Ou, Y. and Tsai, S. (1991) Flavour characteristic compounds found in the essential oil of *Ocimum basilicum* L. with sensory evaluation and statistical analysis. *J. Agric. Food Chem.*, **39**, 939–943.

Sheen, L. and Tsai, S. (1991) Studies on spray-dried microcapsules of ginger, basil, and garlic essential oils. *J. Chin. Agric. Chem. Soc.*, **29**, 226–237.

Sheen, L., Lin, S. and Tsai, S. (1992) Properties of essential oil microcapsules prepared by phase separation/conservation. *J. Chin. Agric. Chem. Soc.*, **30**, 307–320.

Tanaka, T., Hasegawa, A., Yamamoto, S., Aoki, N., Toyazaki, N., Matsuda, Y., Udagawa, S., Sekita, S., Narita, N., Suzuki, M. and Harada, M. (1988) Natural occurrence of aflatoxins in traditional herbal drugs from Indonesia. *Proc. Jpn. Assoc. Mycotoxicol.*, **28**, 33–35.

Tateo, F., Santamaria, L., Bianchi, L. and Bianchi, A. (1989) Basil oil and tarragon oil: composition and genotoxicity evaluation. *J. Essential Oil Res.*, **1**, 111–118.

Yamawaki, K., Morita, N., Murakami, K. and Murata, T. (1993) Contents of ascorbic acid and ascorbate oxidase activity in fresh herbs. *J. Jpn. Soc. Food Sci. Technol.*, **40**, 636–640.

Yan, P. and White, P. (1990) Linalyl acetate and other compounds with related structures as antioxidants in heated soybean oil. *J. Agric. Food Chem.*, **38**, 1904–1908.

INDEX

Other volumes in preparation in Medicinal and Aromatic Plants – Industrial Profiles

Cinnamon and Cassia, edited by P.N. Ravindran and S. Ravindran
Colchicum, edited by V. Šimánek
Curcuma, edited by B.A. Nagasampagi and A.P. Purohit
Eucalyptus, edited by J. Coppen
Ginkgo, edited by T. van Beek
Ginseng, by W. Court
Hypericum, edited by K. Berger Büter and B. Büter
Illicium and Pimpinella, edited by M. Miró Jodral
Kava, edited by Y.N. Singh
Licorice, by L.E. Craker, L. Kapoor and N. Manedov
Piper Nigrum, edited by P.N. Ravindran
Plantago, edited by C. Andary and S. Nishibe
Salvia, edited by S.E. Kintzios
Stevia, edited by A.D. Kinghorn
Tea, edited by Y.S. Zhen
Tilia, edited by K.P. Svoboda and J. Collins
Thymus, edited by W. Letchamo, E. Stahl-Biskup and F. Saez
Trigonella, edited by G.A. Petropoulos
Urtica, by G. Kavalali

This book is part of a series. The publisher will accept continuation orders which may be cancelled at any time and which provide for automatic billing and shipping of each title in the series upon publication. Please write for details.